# Don't Medicate–Educate!

# Don't Medicate–Educate!

## One Family, Three Cases of Autism, Safe Treatment for Dangerous Behavior

Ilana Slaff, MD

ISBN: 0692790209
ISBN 13: 9780692790205
Library of Congress Control Number: 2016917199
Ilana Slaff Medical PLLC Fresh Meadows, NY

# Advance Praise for *Don't Medicate–Educate!*

Here is a brave, wise, and touching book on one of the most complex and devastating disorders out there-autism. Dr. Ilana Slaff's book describes the tortuous journey of her family members with autism and her own struggles to make the world understand their plight. She's clearly a tireless fighter for the rights of those with autism and their long-suffering families. This is a very wise book and is required reading for those whose lives are touched by this condition. She offers wisdom and excellent recommendations on how to cope and utilize resources. I highly recommend this book.

Rakesh Jain, MD, MPH
Clinical Professor
Department of Psychiatry
Texas Tech University School of Medicine
Midland, Texas

As a special-education teacher with twenty years of experience working with students with autism, an adjunct professor at Queens College, a board-certified behavior analyst at the doctoral level, and a certified expert in the field of applied behavior analysis, I heartily endorse Dr. Slaff's insightful description of her journeys encountered through experiences of her twin brothers and daughter, all of whom have been diagnosed with autistic spectrum disorder (ASD). She has identified the elements that have proven to be successful in the treatment of her family members and those that have had little effect. This is an easy read, with many specific examples that are applicable to a multitude of other individuals with ASD.

Carol Fiorile, PhD, BCBA-D, SAS
Thornwood, NY

This was so open, personal, and honest. At times I had tears in my eyes, and at times I laughed. You shared a personal journey of pain, joy, and confusion, but ultimately you educated those around even if they did not respond to you.

I loved how you added your e-mails, which were actually sent by you.

Well done and let's take a step forward to act for each individual as their needs are their own. One size does not fit all, and one treatment does not help all.

Lina Gilic, PhD, BCBA-D
Assistant Professor
Coordinator of Special Education
Department of Educational Specialties
School of Education
St. John's University
Queens, NY

Dr. Slaff deftly describes the trials and tribulations of navigating the many channels needed to secure quality care for people with autism spectrum disorder. She provides a truly unique perspective integrating experiences as a mother, sibling, and psychiatrist. This book provides valuable information for parents and others hoping to understand the protean world of services arranged for people with autism spectrum disorder.

Nathan Blenkush, PhD, BCBA-D
Director of Research
Judge Rotenberg Center
Canton, Massachusetts

This book provides a unique and realistic glimpse into navigating the challenges of autism across generations. Slaff's writing blends humor and history in an insightful storytelling approach that is both thought provoking and deeply moving.

Monica R. Howard, PhD, BCBA-D, LBA
Associate Executive Director
The ELIJA School
Levittown, NY 11756

Dr. Slaff is a tireless, dedicated advocate for her daughter, brothers, and patients. Her open nature allows her to provide a unique perspective from someone who knows autism inside and out.

Lauren A. Goldberg, Esq.
Senior Associate
Kule-Korgood and Associates, PC
Forest Hills, NY

# About the Author

Dr. Ilana Slaff is a sister of identical twin brothers with autism and the mother of a child with autism. She is also a psychiatrist who completed an autism research fellowship at Mount Sinai Medical Center in New York. Dr. Slaff assisted editing the chapter "Complementary and Alternative Therapies for Autism" in the *Clinical Treatment Manual for Autism* (2007) and has been published in the *Journal of Clinical Psychiatry* and the *Psych Congress Network*. Dr. Slaff has given continuing medical education lectures internationally on autism diagnosis and management and has also lectured to families. She has experience in consulting and supervising functional behavior assessments and behavior intervention plans. Dr. Slaff has presented at numerous public hearings, including before the FDA. She has also spoken on New York's *Capitol Pressroom* radio show and appeared on *Your News Now Capital Tonight* for New York.

# This Book Is Not Legal or Medical Advice

All contents of this book including text, images, and graphics are for general informational purposes only. The information in this book is not intended or implied to be a substitute for professional legal advice, medical advice, diagnosis, or treatment. Always seek the advice of your physician or other qualified health care professional. Do not delay seeking medical treatment because of anything you have read in this book. If you need legal advice, consult with an attorney, as this book is not intended to instruct you to take any legal action or inaction.

*To my daughter Batsheva, who always takes care of her sister.*

# Acknowledgments

First and foremost, I want to thank Dr. Matthew L. Israel, the former executive director at the Judge Rotenberg Educational Center, for saving my brother's life. He has also taught me that everyone has potential, and because of him, no matter how hopeless it may seem, I can see a future for my patients. He has shown that psychotropic medication does not have to be the answer and that anyone can learn. No one is untrainable. Unfortunately, throughout history, some individuals who have saved lives using innovative but controversial treatments have been unfairly criticized. Dr. Matthew L. Israel is one among many.

I thank Bill Barnes, from the Clearview School, which my brothers Matthew and Stuart attended. He repeatedly went out of his way to help my family. He called politicians to help stop banning aversives. Years later, he even contacted an attorney to get help for my daughter.

I thank the late Michael Prieto, the lawyer who gave my mother legal advice when he was dying and too ill to make any more court appearances. He also helped a child without a family to pay him.

I also acknowledge the late US district court judge David G. Trager, who issued the court order for New York to fund the Judge Rotenberg Educational Center. Without him, I am not sure Matthew would even be alive today, much less doing as well as he is.

I also thank Eugene R. Curry, a Massachusetts lawyer who was always available, met with the Judge Rotenberg Educational Center families in a snowstorm, and stayed in his office late on Christmas Eve to get documents from my mother for an emergency court motion.

I also want to acknowledge Ralph Antonelli at Judge Rotenberg Educational Center, who truly has been an adopted uncle to Matthew. He started as Matthew's case manager. He has visited the more recent proposed placements with us and has advocated for us whenever needed.

I also thank the new executive director of the Judge Rotenberg Educational Center, Glenda Crookes, who never gives up on her students and is continuing the wonderful work of Dr. Israel.

I thank all the dedicated staff at the Judge Rotenberg Educational Center. Every time I visit the center, I can feel the love and care the staff has for those individuals.

I am grateful to Dr. Carol A. Fiorile, who stuck her neck out to help support Matthew's treatment when other professionals would not and who advocated successfully to keep my brother out of inappropriate placements.

I thank the lawyers for my daughter Talia: Michele Kule-Korgood and Lauren Goldberg. They work late at night in the office making sure children get the help they deserve. I also appreciate their thoughtful review of my book.

I thank the colleagues with whom I worked at Mount Sinai Seaver and the New York Autism Center of Excellence who evaluated Talia and agreed to testify at a hearing when necessary. In particular, I appreciate Dr. Latha V. Soorya, who evaluated Talia and visited on her own time a placement the Department of Education recommended.

I also express gratitude to Debora Thivierge from the ELIJA School, who had the vision to help children with research-based practice. She will do anything for children to succeed. I do not know where Talia and I would be without the ELIJA School. Whenever I visit ELIJA, I can also see the love and care the staffers give to the children.

I also appreciate Dr. Monica R. Howard's review and thoughtful suggestions for my book.

I want to thank all my family members who have been supportive, helped me financially, watched my children, and taken Talia to her therapies. I thank my family members and colleagues who have encouraged me to write this book.

I also want to thank all the other dedicated professionals who helped my brothers and my daughter.

# Contents

# Foreword by Debora Thivierge, BCaBA, CBAA

It's hard to believe that ten years have passed since ELIJA School opened. Back in 2006, it would have been hard - impossible - to predict the long-lasting relationships that we would build with our students' dedicated and determined parents.

While statistics may vary by points, the fact of the matter is that autism predominantly affects males. It was no surprise, then, that ELIJA School's first students were all boys. Talia was our first female student, adding a new layer of diversity to our program. The first time we met beautiful Talia, we immediately noticed her natural curiosity. She looked all around the room. She looked up, and she looked down. However, Talia never once looked at our faces. Beautiful Talia presented herself as a classically impaired child with autism.

Albert and Ilana were the most dedicated parents we had ever met. Ilana shared with us how much time over the years she had spent advocating for her twin brothers (who also have significant developmental disabilities) and simultaneously fighting for her daughter, her Talia. She is an incredible woman fighting a generational battle for her family. It should be noted, too, that Albert and Ilana also are wonderful parents to Talia's sister Batsheva (herself an exceptional young lady).

Ilana and Albert were so knowledgeable about Talia's needs. They had come to ELIJA School seeking effective treatment that would offer Talia a better quality of life. Ilana lived with the fear of her own twin brothers' challenging journeys. She watched everything unfold before her eyes, from their childhood through adulthood. It was clear to me that nothing would ever slow her down or stop her from reaching the goal of securing effective treatment for her daughter.

What I find most inspiring about Ilana is her never-ending fight for transparency in the medical world - to educate herself and help others understand effective behavioral treatment and how prescribed medication is not the only route to deal with challenging behaviors. She opened my eyes as to how sometimes medication can cause more harm than good. Ilana has documented her findings quite succinctly in her various letters and essays.

Ilana fights diligently to spread awareness of alternatives to medication for children and adults with autism. She spends her days and nights writing letters, participating in advocacy groups, and sharing her personal and professional experience as a practicing licensed psychiatrist. Medication abuse has impacted her family so deeply that she has a devastating but valuable perspective to share with the community.

Sadly, Ilana lost her father recently. In the traditional Jewish faith, mourning includes *shiva* - a week for friends and family to pay their respects to the immediate family following the funeral. When I visited the Albert and Ilana's home, I had a few moments to share with Ilana in the reminiscences of her childhood with her father. As I listened intently with all Ilana's extended family sitting around, it was clear how invested her entire family is in Talia's progress. Surrounded by this love, Albert and Ilana's advocacy goes beyond fighting for their daughter. They are here to protect the rights of all people with autism.

When parents come to ELIJA School, we often get to see only one side of the coin. They are desperately trying to find a place for their children to be safe and learn, clawing their way to the light at the end of what may be a very long tunnel. The current New York State educational system is horribly defective, leaving many families with no appropriate options. Parents often have to take a second mortgage on their home and borrow from friends and family to even have a chance to fight for their children to get effective treatment.

After learning about Ilana and Albert's journey, I was stricken with awe at the depth of love that went into their fighting for Talia to get into our program. The vision and intention behind starting the ELIJA School was exactly their battle. We wanted the ability to help families just like Albert and Ilana's and children like Talia.

Often when you interview a family for a private school, you don't have the time to go in depth about where they came from and how they got to your school. The personally written pages that follow tell the emotional struggle of what it was like for Ilana's family and provide a sense of what journey may lie ahead for others. Every family has a different experience with autism; Ilana's is a true example of what a dedicated parent is. Her story will resonate with so many and give them the strength to move forward.

# Introduction

Growing up with identical twin brothers with autism, and then having a daughter with autism, means I have spent most of my life living with someone with autism. It is not a typical life, and I realize other people have their own problems. As a psychiatrist, I can sometimes listen to other people's problems and try to forget my own, at least for a while. However, as a psychiatrist who specializes in autism, sometimes I find some patients get "too close to home." Nevertheless, I enjoy what I do, although I get frustrated-frustrated because individuals are not getting their behavioral and educational needs met.

Many individuals are not receiving the educational therapies they need, including *applied behavior analysis*,[1] so their behaviors are not under control. Many end up needing residential placements and chemical and physical restraints just so no one gets hurt. The chemical restraints, as shown in my brothers' experiences, do not always work to stop the danger. One of the reasons for the system's reliance on chemical restraints is risk aversion: if a patient displays dangerous aggression or suffers a self-injury when behavioral therapies are not available, the doctor can be sued for malpractice for not prescribing medication or other treatment.

What frustrates me the most is how politics and special interest groups maintain the medical and educational system against the families who are too

1 "ABA is a systematic approach for influencing socially important behavior through the identification of reliably related environmental variables and the production of behavior change techniques that make use of those findings." Behavior Analyst Certification Board, accessed January 10, 2016, http://bacb.com/about-behavior-analysis/.

weak or don't know how to fight for their loved ones' rights. I know firsthand it is hard enough just to live with and care for a severely autistic person, a person with dangerous behaviors in your house: how it runs your life, stresses you out, takes away your energy, gives you sleepless nights, and makes you constantly run around. The politicians and self-appointed advocates with agendas may say nice things, pretend to care, and give the appearance they are providing good services, but they may be phony, and appearances are deceiving.

When politics are aligned with special interests, more often than not, they make matters worse. This alignment gets in the way of getting what your child needs, weakens you, ruins you financially, makes you more anxious, and gives you insomnia even on those nights when your child with autism (or other special-needs child) just happens to be sleeping well. This will burn you out.

Either we can be defeated by it, or we can *fight back*, even though we are already going through so much. If we do not fight back, we are allowing politics and other people's agendas to win over our children's lives, and we must never let that happen. There are times I thought I was going to be defeated, that one brother would lose his right to effective treatment, and that he would be left untreated and die from his behavior. Somehow, though, my family always prevailed. And my brother has prevailed, as has his happiness and improved higher functioning.

A lot of thanks have to go to families who advocated for the Education for All Handicapped Children Act (PL 94-142), which passed in 1975. It is only because of those families who came before that we can have funding for the services our children need. It is only because of families today that individuals with disabilities will continue to have their needs met.

In a *New York Times* article in 2002, Joel Levy, then New York City school chancellor, in referring to funding private schools, stated, "You cannot give one kid the Cadillac and the others the back of the bus."[2] Although Mr. Levy was correct that children should not be treated differently based on familial economic status, I am not talking about Cadillacs. I am talking about teaching

---

2 Yilu Zhao, "Rich Disabled Pupils Go to Private Schools at Public Expense, Levy Says," *New York Times*, April 17, 2002, http://www.nytimes.com/2002/04/17/nyregion/rich-disabled-pupils-go-to-private-schools-at-public-expense-levy-says.html.

people with medical conditions to live, learn, and function as independently as possible. Everyone deserves effective treatment, and children who receive effective treatment should not be thought of as having a luxury.

The first four chapters discuss my personal experiences with my family members who have autism and the struggles to obtain effective treatment. The book discusses how one brother has been doing well with effective treatment, while his twin has done poorly. It also discusses effective treatment for my daughter.

To determine if a treatment is effective or dangerous requires research. In chapter 5, I discuss published research on positive-behavior treatments, aversives, and psychotropic medications used to treat individuals with autism or other developmental disabilities. The discussion includes specific risks to individuals with autism and developmental disabilities in general. I also discuss what is actually being implemented in real-life settings and advocating for evidence-based treatment. Recommendations to improve access to effective interventions are discussed in chapter 6. In the appendix, I have included a history of evidence-based treatments utilized in autism to familiarize readers to topics discussed throughout this book.

On a personal note, I also have some autistic features, one of which is impaired facial recognition. In residency training, to compensate for not recognizing my patients, I kept a separate binder with detailed notes on each one of them. Knowing I could not physically recognize them, I would recognize them by reviewing my notes before each encounter. I must have done something right, because my supervisor nominated me for the Pfizer Westchester Medical Center Resident of the Year Award and also recommended me for an elective rotation supervising junior residents. Even if something does not come naturally, it still can be learned or otherwise be compensated so that it does not become a problem.

In this book, which took me more than nine years to write (it was always hard to find time), I talk about coping with living and caring for those with autism and making sure they receive the therapies they need. I talk about the added stress and anxiety, my own coping skills, and of navigating a complex bureaucratic system that is more concerned with political correctness and saving money than caring for those in need. At least there is a system, and it is a system that gives children's rights for which we can advocate.

However, it is far from perfect, and effective evidence-based care for severely impaired individuals under the age of twenty-one is, for the most part, only accessible if the family has considerable liquid assets or is able to secure significant loans. For individuals over the age of twenty-one, with limited exceptions, there is no legal right to care, and therefore, effective treatment for those with severe behaviors is almost nonexistent. This is true as well for individuals with intellectual disabilities. For so many parents, teachers, and therapists, all the hard effort we put into educating individuals (some who regress because there are no appropriate services after the age of twenty-one) becomes for naught. That is what must change!

# One

## My Brothers' First Years

This chapter discusses the initial diagnosis and the quest to find effective treatment. I also discuss my own problems and emotions growing up with siblings with autism.

### Initial Diagnoses, Therapies, and Schools

My brothers Matthew and Stuart are identical twins, born in 1971. They are twenty-six months younger than me. I remember well the struggles my parents went through. My father could not cope with the problems, leaving it to my mother to handle everything. I remember her worrying about finding placements and obtaining funding. The struggles have been there since my brothers were two years old, and they still continue. They will never end, or they will end in death.

My brothers grew up at a time when autism was considered largely the parents' fault. At first the pediatrician denied any problem, despite my mother's concern. My mother finally decided she did not care what the pediatrician said because she knew there were problems. When my brothers were two and a half years old, my parents took them to a clinic on my mother's own volition. At this clinic, one doctor, apparently believing the "refrigerator mother" theory of autism, told my parents during an evaluation that the reason my brothers had these problems was that my parents did not love my brothers and were not affectionate with them. The doctor told my parents they should institutionalize their sons.

Another time an evaluator was working with one of my brothers, and she suddenly said, "He's retarded." She also said, "You shouldn't be surprised. You brought him here." My mother was in utter disbelief, and on the way home, she drove the car over a wooden barrier in the parking lot. At a different appointment, she told the same evaluator that my brother could read. My mother taught all three of us to read starting at the age of one. She taught us so young that I have no recollection of *learning* to read. The evaluator did not believe my mother, but after she insisted, the examiner held up a card with the word "you" in front of my brother, which he read out loud. He could do this even though he could not hold any sort of a conversation or even say anything spontaneously. The examiner actually told my mother she was wrong to teach him reading skills. Another psychologist at the same clinic told my parents to enjoy their children while they can.

My parents took my brothers to the play therapy at this clinic for about two years because nothing else was available. My mother had tried to place them in a nursery school, but they were thrown out after one day. The following school year, another nursery school accepted them, and the owner/director told my mother to register one of them for the three-day session and one for the two-day session. The director told my mother that she thought my brothers might get something out of it. I was told Stuart spent his time rocking back and forth under an easel in the classroom.

My mother finally found something else: Columbia University Teachers College in Manhattan. This was my brother's first special-education school. The nursery school director was happy my mother found this place, released my mother from the tuition contract, and even went into Manhattan to tell the new teacher about my brothers. My mother had to drive my brothers every morning in rush-hour traffic from the Bronx, but it was well worth it. This was in 1975, so there was no busing for children under kindergarten age. It did not make sense for her to go home because the program was only for three hours, so she stayed around the entire time until it was time for them to go home. The school provided funds for my mother to hire someone to ride with my brothers in the back of the car.

The program was a public school program. It was initially federally funded as a pilot program, and later the city took over. It was truly a model program. The head teacher, who was very dedicated, was employed by the New York City Board of Education (BOE). All the paraprofessionals were graduate students at Columbia University Teachers College. There were one-way mirrors for classroom observation by parents, who could watch as long as they liked. There was parent counseling by a psychologist.

Once in a while, my mother would send me to school with my brothers, so I could see what their school was like, and the head teacher always got me involved in the classroom. I would blurt out the answer to questions at circle time, and I still remember her telling me, "I know you know the answer. I need to see if they know it." I would later use the same words in response when I would ask a question to help our older daughter Batsheva with her homework and my husband would inadvertently call out the answer.

The teacher made notebooks for my brothers, and she would write daily in them something they did that day. I once suggested doing the same thing with them on the weekends, and she thought that was a great idea. She made notebooks for me to do with them. I remember Matthew receiving speech therapy through video with Dr. Seymour Rigrodsky, head of the speech department, and how amazed I was when he got my brother to say the word "cup."

One winter break it snowed a lot, and my mother was worried about driving the car to get my brothers to speech therapy. Every day during that holiday break, when speech therapy was available, my mother took the three of us into Manhattan on the bus and subways. My brothers got their therapy. When my brothers were older and were eligible for busing to school, at first the Committee on the Handicapped (COH), which later changed its name to the Committee on Special Education (CSE), tried not to fund their transportation, telling my mother that there were schools in her own neighborhood. She later obtained the busing, because, as she learned from Advocates for Children, my brothers were legally entitled to it because they were already enrolled as students in the program.

The school program was still a half day. After a time, Stuart attended the morning session, and Matthew attended the afternoon session. There were still light moments such as when one day a puzzled bus driver, who took one twin home after the morning session and the other to school for the afternoon session, asked my mother, "Why am I bringing this child home for lunch?"

## Getting Around the "Gag Rule," Board of Education Style

When my brothers were aging out of their preschool program, the head teacher had to write a report that included a recommendation. Being employed by the BOE meant that she could get into trouble if she recommended anything other than a public school placement, even if it was inappropriate. What she did to avoid getting herself into trouble was to document all their needs. She wrote, "I am sure you will find an appropriate placement." This "gag rule" unfortunately continues today, so employees of the New York City Department of Education "cannot" and "will not" recommend placements or services according to their professional judgments but rather are constrained by political bureaucrats who have never met or evaluated a child.

However, my mother already knew that she did not like the autism program at the local public school. When my brothers were about to age out from the preschool at Columbia, the school district violated the Riley Reid law.[3] This law was based on a court decision regarding two handicapped children who were unable to go to school. Their mother was told one of them might need to wait two years for a placement. The judge ruled in favor of the family and ordered that services needed to be provided in a timely manner.

---

3 Riley Reid, a minor under the age of twenty-one years, by his mother, Ellen Hoffman, and Benjamin Kennedy, a minor under the age of twenty-one years, by his mother, Virginia Kennedy, on behalf of themselves and all others similarly situated, plaintiffs-appellants, v. the Board of Education of the City of New York, and Harvey B. Scribner, individually and as chancellor of the Board of Education, defendants-appellees, US Court of Appeals for the Second Circuit, 453 F.2d 238 (2d Cir. 1971), December 14, 1971, http://law.justia.com/cases/federal/appellate-courts/F2/453/238/386024/.

My mother could obtain a private placement for my brothers because the COH would not be able to place them in the required time. The law meant school districts had to follow specific time frames, starting from when both an Individualized Education Program (IEP) is requested and the consent for evaluations is signed by a legal guardian. The district must conduct an evaluation, including an IEP meeting with a special-education teacher present, a general education teacher, if applicable, and a psychologist at the initial meetings, and reevaluations. The district then must decide on a placement. If the COH/CSE is not in compliance, then a child is entitled to a state-approved private placement.

As soon as the timeline expired to obtain a placement, my mother went to the public library and obtained the names and addresses of twenty-one politicians. Sitting for hours at a typewriter, she wrote letters to all of them. Even though the IEP meeting was held after the placement deadline, the COH refused to agree to a private placement. However, a few days after the IEP meeting with the COH, my mother received a phone call from Governor Hugh Carey's secretary, who informed her that she could select a private placement. My mother selected Reece School in Manhattan.

Even when they were older, my mother continued to bring my brothers to Columbia after school for speech therapy. However, every time she parked the car, they got out and ran in different directions, so she requested-and was given-one of the three car-parking spaces by the building's entrance.

## My Early Years

Given the issues with my brothers, my parents had the attitude that I did not have a disability and that I would do fine at any school I attended. This meant that I went to the local public school, which was in a violent neighborhood. In my school, the class size was large, my supplies and my lunch were sometimes stolen, and I was physically abused. Classmates would either ask me if or tell me that my brothers were retarded, which always made me uncomfortable. However, my father was rigid and did not want to move out of the neighborhood.

My parents could not afford to send me to a private school. My mother had been a mathematics teacher and passed three actuary exams, but because of my brothers, she was not working, so we lived on one income. When I was young, I decided I would work hard and get a good job later on so my children could attend a good school.

When I went to visit my brothers' school, I was jealous. I wished I could go to such a nice school. When the problems at my local school became severe, my parents sent me to live with my grandparents in a better neighborhood, so I could attend school, safely.

When my mother had to take my brothers to speech therapy after school, I was in third grade, just seven years old. I would wear a key attached to a string around my neck and be home alone until about six o'clock. My father would call me from work to make sure I locked the front door. I did not mind being home alone, although I did watch too much television. I really never thought this was unusual until Batsheva was about seven or eight years old and I told my husband I planned to leave her alone at home while I went out for a short while. My husband vehemently protested, stating that she was too young.

## Busing Problems

My mother later struggled with a school bus strike that lasted for five months. She had to get my brothers from the Bronx and into Manhattan to Reece School every day in rush-hour traffic to and from school. She also drove another child, Darren, who was on the same bus route but attended a different school. She and Darren's father would take turns driving the children. There was little she could do while sitting in traffic whenever Stuart would say, "Matty hit Darren," or "Darren hit Matty." One time Matthew attacked her while she was driving. My mother pulled the car over to the shoulder and waited for him to calm down. School bus strikes can put people into dangerous situations.

The bus was not perfect either. One time my brother Matthew was physically attacked by another child on the bus. He came home from school highly agitated, and it took an hour for my mother to calm him down. She only found out what caused this attack by speaking with another child, who-unlike my brother-was quite verbal. According to this account, Matthew had torn down the Christmas decorations on the bus, and the bus matron had permitted another student to attack him. The school was defensive, insisting that the child involved would never do such a thing. After my mother complained, the New York City Office of Pupil Transportation met the bus driver at the school the following morning. My brother was never attacked on the bus again. However, on another occasion, the bus returned Matthew to the house, and my mother was told that because of his behavior, they could not take him to school.

When my brothers were babies, they banged their heads against their cribs, so my grandfather padded their cribs. Stuart outgrew banging his head. Matthew did not; in fact, it only got worse over the years. He became aggressive toward others too. For example, he would attack me or anyone else who had the hiccups.

Eventually, in 1982 at the age of ten, Matthew was expelled from Reece School. My mother located another school in Westchester, New York, that would take him: Clearview School. At the IEP meeting, the CSE agreed to this placement. However, no busing was available because the school was in Westchester and we lived in the Bronx; the New York City CSE stated they did not bus outside of the city.

My mother again had to drive Matthew to school and pick him up on a daily basis. Because we had moved, this forced me to start school a week and a half late because my mother could not wait the whole day at the new public school for registration and also handle Matthew's transportation. My mother became quite frustrated from the pressure and accused my father of not handling anything.

My father then called the CSE and inquired about transportation. The CSE asked him if he had a "Nickerson letter," which is a letter New York City parents can request to select a private-school placement if the CSE makes legal errors. My father, never having any involvement with the CSE or an IEP, and not having any idea what a Nickerson letter was, simply stated, "When I see Mr. Nickerson, I'll ask him to send me a letter."

After that, my mother got involved the Advocates for Children to assist her and later became a volunteer advocate for them. New York City finally contracted out with a Westchester bus company to transport my Matthew to school.

## Going to Europe for Help

In the summer of 1982, we went to France for a listening program now known as Auditory Integration Training. The treatment consisted of twenty sessions of music piped in through earphones at various frequencies for twenty to thirty minutes. I was twelve years old. Even though it was a new treatment, and my parents did not know if it would possibly be effective, my parents did not want to miss an opportunity to help my brothers.

On the day we arrived in France, Matthew banged his head into a window in the waiting room of the doctor's office; the window broke.

One day, we spent time by a small zoo near the doctor's office. Matthew did not want to go home and threw a screaming tantrum, kicking my mother. A man approached us, saying, "This is a public place." My mother responded, "Then you get him back to our hotel."

When we finally returned to the hotel room on the third floor, Matthew picked up part of the luggage carrier, threw it from the terrace, and in an instant, took the other part and threw it off too. My mother tried to stop him, but he was too fast. I remember my mother panicking that if someone was hurt, she might get arrested. She debated whether to check if anyone was injured, and finally did so. Nobody was hurt, but people were looking up at the hotel. At other times, Matthew threw tantrums and could not be brought into a restaurant, so my parents took turns eating dinner, while the other would watch Matthew.

After the treatment, we stayed two more weeks in Europe sightseeing. My mother thought it might help my brothers' language development. Because of my brothers' obsessions with riding trains, in Paris we had to take the

Metro to avoid temper tantrums even when riding a bus would be much more convenient.

One time Stuart had to go to the bathroom, and we were by a restroom that required paying a franc to use. Stuart decided he wanted to be alone, so he closed the door. Realizing Stuart could not be trusted to handle himself in a public toilet, my mother opened the door. Stuart pushed her out. This happened a few times before my mother noticed a sign, "One franc every time the door closes." The woman who owned the concession approached my mother and demanded more francs. My mother refused to give the woman any more francs. Meanwhile, Stuart stayed on the john, with intestinal discomfort, too sick to get up. The woman then called the police. The policewoman sized up the situation and walked away.

One day we went to Montmartre. It was a nice day, and we walked to the top of the hill. On the way up, Matthew had some sort of attack. He yelled and banged his head against a rock. My mother comforted him and held him tightly with one hand to prevent him from hurting himself again. She held onto her purse with her other hand because young pickpockets were standing nearby. They actually looked sympathetic, and I am sure at that moment, they could have taken her pocketbook if they really wanted to.

We went to Holland for an international exposition on flowers and gardening called the Floriade. My brother Matthew ran away while we were there, and the police would not help. Although the Floriade was gigantic and had a train that went from section to section, it had only one entrance and exit. My parents split up to look for him, and my mother told me to stay by the exit. While I have no recollection of any of the flowers that day, I still remember standing by the black revolving door wondering how I could possibly stop him if he was determined to leave. My father later found Matthew having a good time playing in the mud.

Traveling with Matthew was nearly impossible. On a trip to Disney World, my parents discovered Matthew banging his head on the medicine cabinet mirror. Local trips were often canceled because of Matthew's behavior: he would be out of control, so we could not go anywhere. As a child, I was frequently deeply disappointed. His behavior interfered so much with our quality of life, and things only became more difficult as he got older.

Having identical twin brothers with autism gave me a lot of responsibility at a young age as well as problems with peers. However, my parents' example also made me patient, persistent, and resilient. I also learned to seek my own answers and not blindly follow others.

# Two

## Matthew-Intensive Positive Reinforcement with Skin Shocks Saving His Life and Yet the Government Trying to Stop His Treatment

In this chapter, I discuss Matthew's life-threatening behavior and treatments. I discuss the serious side effects of psychiatric medications that were not even effective and what treatment was helpful. I further discuss how government officials have unsuccessfully tried to interfere with the treatment.

### Hospitals, Medications, and Surgery

When he was older, Matthew liked to crack his head into sharp edges, including wall corners, a behavior that could have caused him blindness or have given him a deadly infection or stroke. My mother would take him to the hospital at all hours of the night for sutures and x-rays. He would talk about cats having nine lives, which he picked up from a cartoon he watched, and he would state he had eight lives left. He talked about cemeteries. He also rammed his head into my parents' teeth and gave my mother a black eye.

To try to control his behaviors, he was first given Haldol (haloperidol), an antipsychotic medication, which gave him symptoms of neuroleptic malignant syndrome. This is a sometimes deadly condition that manifests as hyperthermia, rigidity, and autonomic dysregulation-which is part of the nervous system. He was then put on high doses of Mellaril (thioridazine) but continued to head bang. We were at the emergency room so often that staff there would wave and say hello to us.

In the summer of 1988, I remember going on a trip to the White Mountains in New Hampshire with my family, and it felt so stigmatizing because he had to wear a large bandage on his head. On that same trip, he banged his head against the inside of our car's windshield. Once he was admitted overnight to Jacobi Hospital in the Bronx for infected sutures from his repeated head banging.

About one month after this trip, he was admitted to Lenox Hill Hospital in Manhattan for five and a half months. He underwent surgery to close his gaping head wound from head banging and received intravenous antibiotics for his head infection. The first time surgery was scheduled, Matthew refused the intravenous line, stating, "No intravenous." He got off the gurney, sat on a bench, and watched others enter the operating room. The surgeon was physically there but could do nothing, and he could not bill for it, either. After the surgery was canceled, Matthew asked, "Intravenous now?" My mother pleaded with the surgeon to reschedule; he appeared to take pity on my mother and agreed to do so. That time, she stayed overnight in the hospital with Matthew the night before and got the intravenous line placed in his room. This time, my brother complied.

Matthew had a drug cocktail in the hospital that included Prolixin (fluphenazine), Artane (trihexyphenidyl), Inderal (propranolol), Ativan (lorazepam), and Klonopin (clonazepam) to control his behavior. He slept most of the day, drooled, became obese, and developed permanent tardive dyskinesia (abnormal involuntary movements). The medication combination also caused dysarthria, a motor speech disorder, which made his speech unintelligible, which may have caused him frustration and stimulated dangerous behavior. In the long term, his voice has remained high pitched, and his growth has remained stunted. Before he was on these medications, he was taller than Stuart, but since being on psychiatric drugs, he has been shorter.

Despite being on all these medications and having a hospital staff member present with him around the clock, he continued to need repeated suturing for his head banging. The staff tried to accommodate his every desire, treating him like a king to avoid his behaviors, such as when he ran into a quarantined area. They kept all his favorite Jell-O flavors in a refrigerator in the room for him. Except for one night, he always had his own room because he could be dangerous to a roommate. When he showed extreme agitation, female aides would run out of the room for help. My mother requested male aides and was told it was not possible-until she complained to the patient advocate.

At one point, my mother became concerned because she could not find any facility for him. The insurance company threatened not to pay Lenox Hill Hospital because it was not a psychiatric hospital, even though no psychiatric hospital would accept him. My parents were told they could be held responsible for all hospital costs. I remember my father wondering if the hospital could take our house. However, a sympathetic lady at the insurance company finally authorized payment. His hospital bill for that period was more than $90,000, exclusive of the 1:1 staffing, which the insurance company would not reimburse. Looking back, I do not know how they did it, but almost every day either my mother or my father went to visit him via the subway, despite the fact they had another son with autism living at home.

## Behavior Therapy: Safe and Effective Treatment

Matthew received no education in the hospital because he attacked the schoolteacher. The New York City Board of Education (BOE) told my mother that no educational facility would accept him, and they asked her to waive his right to an education. However, through her physician my mother had heard about the Behavior Research Institute (BRI), which in 1994 became the Judge Rotenberg Center (JRC), located in Massachusetts.

Before starting at BRI, my mother was well informed about BRI's use of aversives. Aversives are procedures that are arranged as consequences for undesired behaviors with the purpose of decreasing the future frequency of those behaviors. Some aversives, such as bad grades, money fines, or critical comments, do not involve discomfort or pain. Those aversives are often very effective with most persons and are widely used and accepted procedures. Unfortunately, the reality of some situations is that they are sometimes ineffective with individuals who have autism. For these individuals, effective aversives may have to include an element of discomfort or pain, such as a spray of water mist to the face, a spank, or a pinch, in order to be effective. At BRI one of the principal aversives employed was a two-second shock to the surface of the skin of an arm or leg by a medical device that has been registered with the Food and Drug Administration. The device is operated by a staff using a remote control.

All aversives at BRI were employed only after an adversarial hearing before a Probate Court judge, in which an attorney was appointed to represent Matthew's interests, as distinct from the interests of BRI. The judge had to decide, first of all, whether Matthew was competent to make his own medical decisions. If not, the judge then decided whether, if Matthew *were* competent, he would have chosen to a treatment that included the use of aversives. The first judge who entertained such a hearing (which, in Massachusetts prior to 1987, had only been used when treatment involved the use of antipsychotic medication) was Chief Judge Ernest I. Rotenberg. After he died in 1992, the school was renamed in his memory.

My parents had to sign consent forms for aversives prior to Matthew's arrival. However, the school's practice was to try using a powerful scheme of rewards-only for desired behaviors for a substantial period of time-for over a year in the average case-before supplementing the rewards systems with the use of aversives. My parents also had to fill out multiple-page forms to assess the things Matthew liked, so the school could consider using them as rewards (technically, *positive reinforcers*). The school would not add the use of supplementary aversives until it was clear that rewards alone were insufficiently effective, and until the school received court permission as described above.

Aversives have been used inappropriately and unethically in the past. This was used in the past as a treatment for homosexuality.[4] Aversives have also been used for punishment for criminal behaviors, such as illicit drug use and

4 Michael King, Glenn Smith, and Annie Bartlett, "Treatments of Homosexuality in Britain since the 1950s- an Oral History: the Experience of Professionals," *British Medical Journal* 328, no. 7437 (February 21, 2004): 427, doi:10.1136/bmj.37984.442419.EE.

sexually inappropriate behavior.[5] However, aversives are effective when combined with dense positive reinforcement, and aversives without dense positive reinforcement can increase agitation.[6] There is a literature analysis printed in the National Institutes of Health (NIH) Consensus Development Conference on the Treatment of Destructive Behaviors, which involved seventeen shock studies and included twenty-two subjects with intellectual disability and/or autism with self-injurious behavior. The self-injurious behavior included head banging, pinching and biting self, hitting self in multiple areas, eye poking, pulling out hair and skin, chronic vomiting, and rumination. These behaviors resulted in severe tissue injury, a detached retina, dehydration, weight loss, and emaciation. Eighty-six percent of subjects showed a 90 percent or greater suppression of their behavior, and 80 percent showed 100 percent suppression of the behavior. One hundred percent of the cases with 90 percent or greater suppression of the behavior showed improvement in one to five days, and the maximum effect was obtained in one to ten days. The follow-up varied. However, individuals who showed 100 percent suppression initially continued to do so at the follow-up.[7] With Matthew, aversives have always been used solely for his dangerous and even life-threatening behavior and in combination with highly dense positive reinforcement.

On a wintery day in 1989, I, along with my uncle, went with my mother when she brought Matthew to BRI. In a taped interview, my mother was asked what she wanted for my brother. My brother asked to go back to the hospital, and she told him to behave. My mother told him that he was a boy and that boys went to school.

---

5 Council on Scientific Affairs, Council Report, "Aversion Therapy," *Journal of the American Medical Association* 258, no. 18 (November 13, 1987): 2564.

6 Dorothea C. Lerman and Christina M. Vorndran, "On the Status of Knowledge for Using Punishment: Implications for Treating Behavior Disorders," *Journal of Applied Behavior Analysis* 35, no. 4 (2002): 455, doi:10.1901/jaba.2002.35-431.

7 Michael F. Cataldo, "The Effects of Punishment and Other Behavior Reducing Procedures on the Destructive Behaviors of Persons with Developmental Disabilities," in National Institutes of Health the Consensus Development Conference Statement, "Treatment of Destructive Behaviors in Persons with Developmental Disabilities," *U.S. Department of Health and Human Services, National Institutes of Health, NIH Consensus Development Program, Office of Disease Prevention, Consensus Development Statement*, September 11-13, 1989, 231-341.

I remember a school staff member approaching my brother with a toy school bus and explaining that he was going to take a school bus soon. Matthew then noticed *Wheel of Fortune* playing on a television in another room, so the staff member took him to watch the show. My mother, uncle, and I left. Matthew has been at that school ever since.

At BRI, Matthew was weaned off his psychotropic medications. He was lively, and he talked again. A few months later, when we asked him if he wanted to come home and visit, he repeated over and over, "I stay at BRI."

The program has always been very individualized. When he first arrived, my mother would ask him daily what he had for dinner. He always replied, "Eggplant parmesan." When my mother inquired about this answer, the staff explained that, on every day he behaved, at four o'clock he would receive a small portion of eggplant parmesan. He has had many other simultaneous positive-behavior contracts to earn preferred items or activities multiple times a day as well as overnight and multiday contracts to obtain reinforcers of his choosing. For example, he would be excited about daily trips to the reward store if he behaved, where he would spend his tokens on snacks or fun activities. In the morning, if he was ready quickly for school, he earned time to watch cartoons. If he passed his multiday contracts, he would earn trips to his favorite places.

Not long after his arrival, despite the intensive positive reinforcement, Matthew continued his life-threatening self-injury and aggression. The school then decided it was time to supplement its reward program with the use of aversives.

> These included a two-second skin-shock, known as the Graduated Electron Decelerator (GED) which he received an average of less than once a month to control his behavior. To experience what my brother feels, I have tried this on myself. There was some pain but it was completely gone after two seconds. There are 119 peer-reviewed articles

that support the use of skin-shock some of which are discussed elsewhere in this book.[8]

After Matthew engaged in a behavior that would lead to head banging, such as attempting to go to an unsupervised area where he knew he could engage in self-injury without interference, he received a skin shock. To further prevent head banging, he received applications for less intensive self-injury, such as after Matthew would rub skin forcefully against objects, picking his nose to bleed or bite his lip. The skin shock would deter him from engaging in further self-injury. These treatments completely stopped his head banging, although he still scratched and pinched himself at times to make himself bleed. His health-dangerous behaviors originally occurred four hundred to five hundred times a month. With intensive positive-behavior therapy, the frequency of his health-dangerous behaviors decreased to sixty to seventy a month. With the skin-shock treatment, the self-injurious behaviors decreased to never more than forty a month-most months he had less than ten episodes, and many months he had no self-destructive behaviors at all. His physical aggression on arrival to the school was nine hundred to thousand times a month. This decreased to four hundred to five hundred times a month with intensive positive-behavior therapy. After adding the GED, the frequency dropped to zero to forty occurrences a month with most months at zero. Property destruction on arrival occurred eighty to ninety times a month. This decreased to fifty to sixty times a month with positive treatment only. After skin shock started, the frequency was zero to two a month.

Besides my brother, the JRC has published research on its successful treatment of life-threatening behaviors with supplementary skin shock. In one study involving seven individuals, which was conducted at the JRC, life-threatening behaviors were successfully treated with the use of supplementary skin-shock treatment after the students were expelled from other programs, some of which employed skilled behavior analysts and used nationally regarded consultants. For one student in particular, "aggression and self-abuse were so frequent that he had to be restrained…'more than 70' times per week," according

8 Bibliography on Skin-Shock, accessed January 24, 2016, http://www.effectivetreatment.org.

to the prior school's discharge summary, "and each restraint required up to five teachers. His self-injurious behaviors included 'body hits to the environment, head hits to wall and floor, body punches, face or head hits, self-bites' and hand contortions (intense wringing of hands and fingers). These behaviors caused 'bruises, scratches, swelling of joints, cuts to forehead caused by intense head-to-floors (while wearing a protective helmet)...[and] fractured bones in his hands on two occasions.' Property destruction included throwing objects, ripping materials, turning over furniture, and throwing large heavy objects. His aggressive behaviors, such as hitting his mother while she was driving, prevented him from having any home visits and curtailed community outings from School A."[9] His progress is displayed in the graph below.[10]

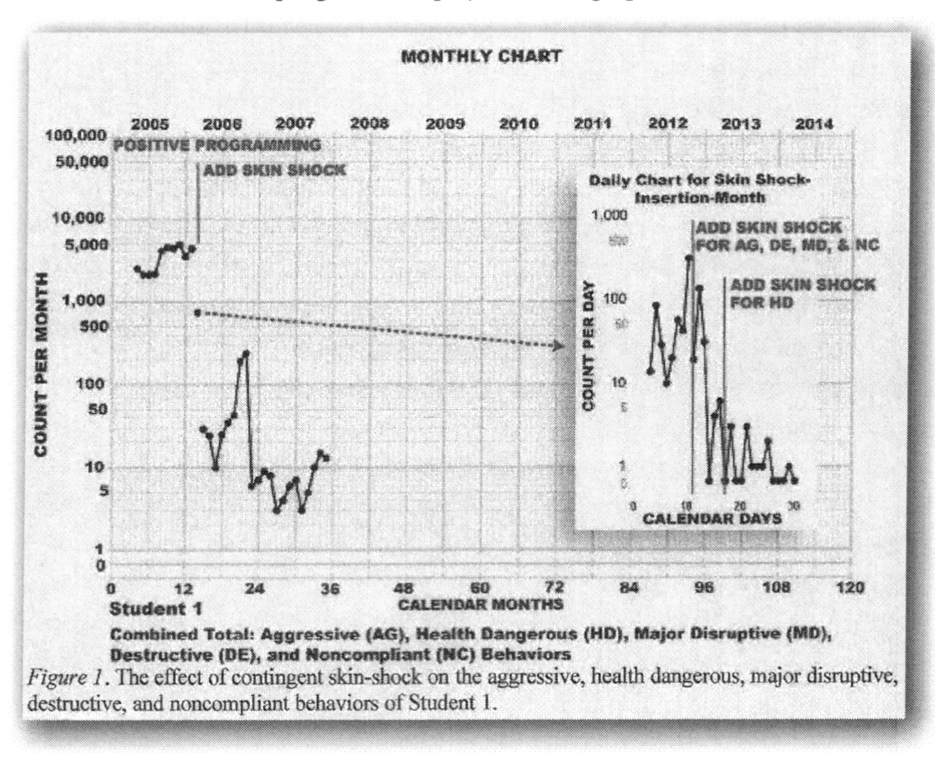

*Figure 1*. The effect of contingent skin-shock on the aggressive, health dangerous, major disruptive, destructive, and noncompliant behaviors of Student 1.

9 Matthew L. Israel, Nathan A. Blenkush, Robert E. von Heyn, and Christine C. Sands, "Seven Case Studies of Individuals Expelled from Positive-Only Programs," *Journal of Behavior Analysis of Offender and Victim Treatment and Prevention* 2, no. 1 (2010): 20-21, doi:10.1037/h0100470.
10 Ibid., 22.

Research is further discussed in chapter 5.

Eventually, Matthew was weaned off the skin-shock therapy, and the intensive positive-intervention program continued. However, on September 22, 2011, Matthew engaged in self-injury by puncturing his eardrums. He had to go to the hospital for the first time in more than twenty years. The skin-shock treatment was reinstated. As of October 2016, Matthew has not received a skin-shock application in over four years, but sometimes after a dangerous behavior, he has had to wear the device to prevent a further escalation.

In summary, Matthew has been at the JRC for more than twenty-six years and is doing quite well on no psychotropic medication. We have been able to take him with us to restaurants, museums, and even on trips to other states. He has a paid job in the school sweeping floors. He has attended several family events. Years ago, when he was told he might be returning to New York, he again started to talk about cemeteries and pointed them out on a map.

If Matthew had continued in his head-banging behavior, he could have suffered brain damage, blindness, septicemia, and even death. The skin-shock therapy stopped that behavior. My family unequivocally feels the skin shocks saved Matthew's life.

## New York State Tries to Terminate Matthew's Funding

After Matthew turned twenty-one, his educational funding was no longer available through the (BOE) but rather the New York State Office of Mental Retardation and Developmental Disabilities (OMRDD). This was a complicated process.

My mother tried to bring the JRC program to New York State. She found a woman who said she was a consultant with important contacts at the OMRDD. She took my mother out to lunch at a Hilton Hotel, and then she arranged a meeting with the commissioner of the OMRDD, Elin Howe, who agreed to start a group home in New York with Dr. Matthew Israel, the previous director of the JRC. Howe later became the commissioner of the

Massachusetts Department of Developmental Services, where she authorized regulations banning the use of aversives.[11]

My mother was allowed to form an incorporated agency to start the group home, and the woman with the connections was hired as a consultant. The consultant wanted to sell the state a particular property that would become the group home, but it was in a remote area, and Dr. Israel worried he would have difficulty finding staff to work there. When she could not sell the property, the consultant set up a meeting with a real estate agent in the Bronx but then did not show up to the meeting. She never returned my mother's phone calls.

The New York State OMRDD has tried to stop funding Matthew's JRC placement. When he turned twenty-one, his funding was 60 percent through the state and 40 percent through the city. The Transitional Care law that applied stated the city "may" pay 40 percent, and then the state would be obligated to pay 60 percent.

Initially, there was no problem, and the city and state provided the funding. However, after Mayor Rudolph Giuliani took office, his office sent out a letter in mid-December 1994 stating that the city would cease its funding in two weeks and that placement was not the city's responsibility. My mother and other parents were in a panic. For some unknown reason, the city sent out a new letter that stated there was "good news": funding would be extended for six more months.

The state has tried to transfer Matthew to a facility called the Developmental Disabilities Institute, located on Long Island, and gave my parents an Individualized Service Plan. The "individualized" service plan referenced Matthew as "he/she" and "his/her." Matthew's tested IQ was about forty-eight, yet one sentence of the "individualized" plan stated, "This training may be through continuing education courses at a local high school, classes at local colleges, or vocational training and trade schools." There were no specific interventions stated to target his self-injurious behavior or aggression. The Individualized Service Plan appeared to be a form and not tailored to his specific needs.

11 JRC is currently contesting this ban in court in Massachusetts.

My parents hired a lawyer, and on June 8, 1995, they had a "due process hearing." There was no independent judge-like there is in an impartial hearing for people under the age of twenty-one in educational placements. Instead, to make this decision, there were three panelists employed by the OMRDD (a psychologist, a certified social worker, and a registered nurse) who determined that the placement in his "Individualized" Service Plan was appropriate to his needs. Even this "kangaroo-court" procedure was denied to other parents, as OMRDD "notified parents that it had eliminated such hearings."[12]

Years later, I was the mental-health professional representative at an Individualized Service Plan meeting. The psychologist, who was not present at the meeting, wrote in the report that the individual had no side effects from an antipsychotic medication. It is inappropriate for any psychologist to make such an assessment because psychologists do not have the medical background with which to make such assessments. I reviewed the medical history and physical and realized his assessment was incorrect. We made a team decision that the psychologist had to amend his evaluation and to consider a medication reduction. This psychologist happened to be the same person who voted against Matthew's treatment in that "kangaroo-court" "due process hearing." During that meeting, although what was most important was pointing out the proper medication assessment for the individual, it felt good to be in the driver's seat for a change, and in at least one instance, stop this psychologist from harming another person's future.

## It Is My Belief That Massachusetts and New York Conspired to Destroy the Judge Rotenberg Center

Here is my evidence.

The JRC sued the Massachusetts Department of Mental Retardation (DMR) for violating a previous settlement agreement allowing for court ordered use of aversives as part of a treatment plan. The suit also alleged that the DMR disrupted its relationships with funding agencies and clients. Its

12 Lisa W. Foderaro, "When Not Just Any Home Will Do," *New York Times*, November 27, 1995, http://www.nytimes.com/1995/11/27/nyregion/when-not-just-any-home-will-do.html.

commissioner, Philip Campbell, was found to have lied under oath. Judge L. Elizabeth O'Neill LaStaiti stated the following in her decision on October 6, 1995:

> After taking no action for more than two years on JRC's application for recertification (submitted in July 1991,) Commissioner Campbell in his August 6 letter, implied that JRC had failed to make its case for certification and granted JRC only twenty-five days of conditional certification. The letter also implied that JRC failed to respond to the Department's requests for information. In fact, JRC had fully complied with all of DMR's requests for information. The Commissioner's letter falsely implies to the contrary.[13]
>
> There were numerous documents which were marked as exhibits which demonstrated the falsity of the Commissioner's testimony. These exhibits were in the possession of DMR counsel and known by DMR counsel[14]...This Court found most of DMR's witnesses not to be credible, and the Court felt the need on numerous occasions, after repeated instances of contradictory sworn testimony, to remind witnesses that they were under oath and had to tell the truth.[15]

The judge found the DMR was not acting properly, and she appointed an independent receiver for JRC to oversee admissions.

Appearing not to be a coincidence, just before the OMRDD cut off funding from the JRC, the Massachusetts DMR instituted temporary regulations that banned continued funding for students who were initially funded from out-of-state placements if their funding was cut off. Our lawyer in Massachusetts discovered communication between the New York State OMRDD and the

13 Judge Rotenberg Educational Center, Inc. v. Commissioner of the Department of Mental Retardation (October 6, 1995), docket 86E-0018-G1, findings of fact, n.75, http://www.clearinghouse.net/chDocs/public/ID-MA-0002-0002.pdf.

14 Ibid., corollary findings improper conduct by DMR and its attorneys, n.10.

15 Ibid., findings of fact, n.75, corollary findings improper conduct by DMR and its attorneys, n.22.

Massachusetts DMR before the temporary regulations were instated, and this communication was presented in court.[16]

Four families from the JRC, including ours, organized to sue the OMRDD. In this process, the families went through four lawyers, and each family spent more than $10,000. Originally, more families were involved, but several agreed to in-state placements and never pursued the case beyond the first of the four attorneys. The lawsuit also initially involved other schools, but because we allowed our children to receive aversive treatments, the other families involved were not interested in working with us and used a different attorney.

We started at the state court, lost, and moved to federal court. The first lawyer, a known special-education attorney, would not pursue the case past the "kangaroo-court" proceeding discussed above. New York state senator Carl Kruger contacted the lawyer who represented the other schools, which is how we found our second lawyer, Michael Prieto.

My mother asked Michael Prieto if he would represent another JRC student who had no family to pay his legal fees. Michael said he would absolutely do so. He even went to Massachusetts to visit the students he represented and said that he prayed for them. He later died during the case. When he was too ill to go to work anymore, he consulted with my mother over the phone and told her how to proceed with the case.

The third lawyer we used missed a court appearance, so the families filed a grievance with the local bar association, had a proceeding, and received a partial refund. During the time we were without a lawyer, my mother represented Matthew herself in federal court. One of the state's attorneys laughed in front of the parents, saying there was no way they could win the case.

The federal court proceedings for my brother and others stated the following:

> A final communication subsequent to the cessation of city funding was a June 27, 1995 letter from Governor Pataki to a TCF [Transitional Care Funding] recipient's parents. It is indicative of the conflicting

16 Ibid., findings of fact, n.219-21.

messages that were sent to parents/guardians by State officials. The letter stated:

> The primary goal of the legislation is to provide appropriate services to children and young adults in New York rather than in out-of-state placements. To this end, individuals who have aged out, or will soon age out, of their child care or educational placements are to be offered appropriate placements, assimilating into the adult care system operated by the Office of Mental Retardation and Developmental Disabilities (OMRDD) in New York State.
>
> My staff has conferred with Commissioner Maul on this issue, and I am confident that OMRDD is sensitive to each individual's circumstances and needs.
>
> Individuals and their families may also benefit from the due process hearing procedures which provide a more formal opportunity to express their concerns over the proposed placements.

Pltf.Mot.Ex. 2, Ltr. dated 6/27/95 Pataki to Resnicks. Despite the governor's statement, which was presumably based on information provided to him by OMRDD Commissioner Maul or his staff that "due process hearings" would be provided, only two weeks later OMRDD notified TCF parents/guardians that it had eliminated these due process hearings.[17]

On July 7, 1995, the New York Office of Mental Health (OMH) sent police officers to the JRC-without notifying the parents-to transport a student to Elmira Psychiatric Center. This incident occurred even after the school informed Lewis Campbell, an OMH official, that the school had contacted

17 Brooks v. Pataki, 908 F. Supp. 1148, United States District Court, Eastern District of New York, November 16, 1995, docket 95-2902, background section G, https://casetext.com/case/brooks-v-pataki.

student's father, who would not consent. The school refused to put the student on the bus. New York State attempted to do the same with different students at different schools.

The families finally won in the federal district court. Judge David Trager said that when the state sent a bus to collect the students, the state was taking responsibility for them. He also wrote, "The State's refusal to fund out-of-state placements, in view of its failure to complete placement of all TCF recipients, suggests a perverse game of musical chairs where not one but many persons are left standing when the music stops."[18]

After we won in federal court, the state appealed, and we lost two-to-one in federal appeals court. The judges ruled there was no obligation to give appropriate treatment unless a person was incarcerated against their will. The appellate court reasoned as follows:

> The Supreme Court did acknowledge that due process obligation created by cases like Youngberg, which involved the security of involuntarily committed mental patients, or Estelle vs. Gamble, 429 U.S. 97, 97 S.Ct. 285, 50 L.Ed.2d 251 (1976), which involved the provision of medical care to incarcerated prisoners, was still valid, but only as an exception that applied to "certain limited circumstances." DeShaney, 489 U.S. at 198, 109 S.Ct. at 1004. The Court explained that these cases "stand only for the proposition that when the State takes a person into its custody and holds him there against his will, the constitution imposes upon it a corresponding duty to assume some responsibility for his safety and general well-being[19]...[p]laintiffs here are under no state-imposed restraint.[20]

18 Roger Field, "Judge: State Must Pay for Hosp Care of Mentally Ill," *New York Post*, November 17, 1995, 24.

19 Brooks v. Giuliani, 84 F.3d 1454 US Court of Appeals for the Second Circuit no. 1421-August term 1995 (argued January 23, 1996, decided May 31, 1996), docket 95-9178, no. 77, http://openjurist.org/84/f3d/1454/brooks-v-w-giuliani.

20 Ibid., no. 80.

We tried to get an en banc and even appealed to the US Supreme Court, but the case was not accepted.

I once got Mayor Rudolph Giuliani on the phone during a radio show to ask him questions. I was driving in my car and tuned in to the middle of the show. I parked my car and went to a pay phone. At first, I got a busy signal, but eventually I got through. The questions were screened, and the screener initially doubted my story. Finally, I was put through. I asked the mayor why he cut my brother's funding. He told me it was the state that cut the funding. I then clarified the situation: he had first cut the city's 40 percent share. He then inquired about the court case. When I told him that we won in the lower court at the federal level but then lost in a two-to-one appeals decision, he answered that I must be wrong.

Ultimately, the state settled, and Matthew is still at the JRC today.

Appointed judges can make life-and-death decisions. The US Supreme Court has ruled in favor of or against children with special needs at different times and has split decisions in cases such as the issue of burden of proof, with the majority deciding that parents have the burden of proof when challenging a school district individualized education plan that they feel is inappropriate.[21]

During the litigation, my mother attended an OMRDD conference where Commissioner Thomas Maul was speaking. During the question-and-answer period with the audience, my mother was selected. She asked about two students who were sent back to New York from the JRC in Massachusetts and who died from their behaviors. She expressed her concern about her son being brought back. One of these students had a history of elopement (running away) and, at the age of twenty-four, eloped on his bicycle from the Developmental Disabilities Institute-the same placement recommended for Matthew. The individual who eloped "slammed into the side of a water delivery truck" and "died of head injuries."[22]

The other student my mother mentioned also came back to New York and died from medical consequences secondary to self-abusive scratching,

21 Linda Greenhouse, "Parents Carry Burden of Proof in School Cases, Court Rules," *New York Times,* November 15, 2005, http://www.nytimes.com/2005/11/15/politics/15scotus.html?pagewanted=all&_r=0.

22 Collin Nash, "Bicyclists, Pedestrian Killed in 3 Accidents," *Newsday*, September 21, 1996, 12.

which caused blood and bone infections, leading to paralysis and death. He was twenty-five years old.[23]

After this incident, my mother was removed from the conference mailing list, and she received a letter from an OMRDD attorney stating she should not discuss her case publically because it was in litigation. My mother read the letter and said, "I don't have to listen to their lawyer."

I tried to ask a question at a later conference, but they no longer selected people who raised their hands in the audience. The few people brought to the podium to ask questions appeared to be preselected.

Another individual not mentioned at the meeting died one year prior to the former JRC student mentioned earlier who also lived at the Developmental Disabilities Institute. This person also eloped from the Developmental Disabilities Institute as discussed in the following article:

> "The institute had no fences, no gates, no security and there were statements that one or more of its employees were high on drugs that evening," Langer [the lawyer for the family] told The Post...Michael Darcy, the director of the institute, confirmed the grounds have no fences or gates.[24]

After my mother refused the Developmental Disabilities Institute placement, the state wanted to place my brother and other JRC students in a special wing at the Bernard Fineson Developmental Center in Queens. During a visit to the facility, my mother noticed extremely bad mattresses. She told the facility that she wanted good mattresses. When she toured the facility a second time, the staff was quick to point out the quality of the mattresses. Regarding the Bernard Fineson Developmental Center, we had two meetings with Dr. Richard M. Foxx, a consultant the state

---

23 N. R. Kleinfield, "James Velez, 25; Tormented by Baffling Illness," *New York Times*, October 29, 1999, http://www.nytimes.com/1999/10/29/nyregion/james-velez-25-tormented-by-baffling-illness.html.

24 Roger Field, "Dead Autistic Boy's Kin Suing Long Island Institute," *New York Post*, September 3, 1995, 19.

hired. As the sister of Matthew and as a medical professional, in 1996 or 1997, I attended one of the meetings at which Dr. Foxx, Dr. Beth Mount, and OMRDD associate commissioner Katherine Broderick were present. We were told the program would not include aversives, and although Dr. Foxx tried to endorse this nonaversive program to us, Dr. Foxx wrote a chapter in a textbook published in 2005 titled, *Severe Aggressive and Self-Destructive Behavior: The Myth of the Nonaversive Treatment of Severe Behavior.*[25] Dr. Foxx would not sign the positive-only treatment plan, although an OMRDD psychologist who was not present at this meeting signed off on such a treatment plan.

I said I wanted to see research studies that found that using a positive-only treatment plan would always be effective to treat life-threatening behaviors. I also told Dr. Foxx that I was a doctor and that when I wrote treatment plans, I signed them, so I wanted to know why he did not sign the treatment plan.

Initially, he replied he did not sign the treatment plan because he was only a consultant. He then stated he would sign it, but he never did. He was also just consulting for a few days and would not be around to monitor the program.

I also received a handout on "mapping," a person-centered planning approach created by Dr. Mount, who received her PhD not in psychology or behavior analysis but rather in organizational development and political science. I asked, "What is the scientific efficacy of this mapping?" Confused, Associate Commissioner Katherine Broderick repeated the question back to me. I clarified that I wanted to know what published research studies showed evidence of the effectiveness of "mapping" for dangerous behavior. She said she would send me information, but she never did. During this verbal exchange, Dr. Mount remained silent. It is no wonder she did, as her person-centered approach lacks accountability, as explained elsewhere in this book.

---

25 Richard M. Foxx, "Severe Aggressive and Self-Destructive Behavior: The Myth of the Nonaversive Treatment of Severe Behavior," in *Controversial Therapies for Developmental Disabilities: Fad, Fashion, and Science in Professional Practice*, ed. John W. Jacobson, Richard M. Foxx, and James A. Mulick (Mahwah, NJ: Lawrence Erlbaum Associates, 2005), 295-310.

## After Twenty-Five Years at the Judge Rotenberg Center, New York State Tries to Relocate Matthew Again

Again, more than fifteen years later, the Office for People with Developmental Disabilities (OPWDD), the new name for OMRDD, tried to place Matthew back in New York. Numerous not-for-profit agencies had visited him and other students but rejected my brother after finding out his history. Meanwhile, the JRC still has not been allowed to open a program in New York.

However, on January 2, 2013, when I called my mother to see if she could pick something up for me, she was upset and told me she just received a letter from Opengate, an agency in Westchester interested in accepting Matthew. Aversives are no longer permitted in OPWDD-funded agencies in New York and are now an outright violation of their regulations. My mother and I both were concerned that the placement would possibly be unable to keep him safe, and we did not want to see him suffer from toxic drug treatments, but we would visit before making any conclusions. She told me she was faxing the letter to our lawyer, and I told her to contact the JRC.

Opengate called my mother while we were taking Matthew to the Boston Science Museum. My mother listened to her voice mail and became tense. I told her to relax and enjoy the museum and that there was nothing that could be done right now.

The next day, Opengate's director of community services, a behavior analyst or specialist, spoke with us by phone. The director was a social worker. The behavior analyst or specialist was not board certified in behavior analysis, and her credentials were just a master's degree in clinical counseling psychology. We were not impressed.

We tried to find our own expert witness to evaluate Matthew and to visit Opengate, in case we were able to have a hearing if we did not agree with the placement. I was so frustrated finding one, because many professionals refused to get involved with the JRC. I was feeling down, stressed, and anxious, and had nightmares. I believed these people did not care about my brother's life. I wanted to challenge these professionals-who were against the JRC-to open their own facility to take care of people with life-threatening behaviors, but without medications with life-threatening side effects.

I was wondering how I could function at work like this, but one Monday morning, I found a willing expert. She asked me about Matthew's history and told me she would do the evaluation. I fell behind with my work because I was on the phone with her, but I was much calmer and able to focus. After she visited Opengate with us, I ran into her again at my daughter's school, where she was also evaluating my daughter. I told her how grateful I was she was helping with Matthew, how I had so much trouble finding someone, and how I was so worried about his life.

Here are my notes on the visit to Opengate on February 27, 2013:

> There was the licensed social worker who was the director of clinical services, a psychiatrist and a chief compliance officer, but the sign on his office stated "behavior specialist." Later we visited day programs and met with the director of community services.
>
> Behavior plans were supervised by licensed social workers, and there were two master's degree psychologists (one full time and one twenty hours a week) and one doctoral psychologist who was only there one day a week, all of whom were unlicensed. Staffers were not trained in research-validated and reliable methods in functional behavior assessment or functional experimental analysis. They said they did not use any aversives and later, according to the director of community services, did not do an intensive positive reinforcement intervention with multiple contracts for different reinforcers as was in Matthew's current behavior plan, and they would meet at thirty days to formulate a plan (he could be dead by then). They asked about Matthew wearing a helmet (and they also said sometimes they used arm guards without restricting the ability to move), admitted they would probably initially have to do a lot of physical restraints and also, if dangerous to property, would have to limit where he could go in the home. We stated that in the past Matthew would take the

helmet off and throw it. The expert that we hired pointed out that with the restraints and keeping him in certain areas, he would be in a more restricted environment than he currently was as now he only received skin shock on average of less than once a month. She also pointed out that it was unethical to remove an effective treatment.

[Actually, I later found out he had not received skin shock for about a year.]

We also pointed out that Matthew was opportunistic and engaged in behavior (hurting or biting himself) when he thought no one was watching him. At night, the staff stated some individuals were on every ten-minute checks, but only continuous eyes temporarily, meaning no one would be watching him constantly for more than a limited time period. This would be a concern for Matthew. Later during a tour of the campus residence, [a JRC staff] told the social worker that at JRC Matthew would have to keep his hands above the blankets at all times for his safety. There were two to three overnight staff for six to ten individuals with no cameras. The social worker informed us that New York State did not permit cameras in an intermediate care facility residence. We pointed out Matthew could need three people to take him down.

There were four group homes plus a day program at a separate location and three combined residences with a day program on campus, and it was not clear where Matthew would live, as the social worker said they would have to purchase a house. Staff members said Matthew might do better when he got to know the staff, but they admitted there was high turnover. The new staffers had a one-week orientation and were required to have a high school diploma. They were trained in physical restraint (Strategies for Crisis Intervention and Prevention; SCIP) and CPR. Staffers were not unionized, and the social worker said the pay was not good. (Getting to know people in itself is not an effective behavior program, and the expert we hired asked the staff from JRC if Matthew also got to know staff there too, and of course he did.) When our expert asked if the windows were shatterproof, the social worker was not sure as Matthew had put head through windows, and the medicine cabinet in a residence bathroom was made of glass.

Ninety-five percent of people there were on medications, over half on antipsychotics and some on multiple combinations. When I discussed the side effects of antipsychotics that I was concerned about, such as diabetes, initially the psychiatrist stated she tried to avoid antipsychotics and preferred mood stabilizers. I then pointed out that Depakote (divalproex sodium), a mood stabilizer, for example, caused weight gain, which could bring on diabetes and also caused liver failure. Finally, she admitted over half were on antipsychotics when directly questioned. When I asked what they did if the family refused meds, the psychiatrist stated that she considered the patients there like her own children and she would hope I would agree with her on medications. [I was particularly disgusted when the psychiatrist told me she considered her patients like her own children. I would never say this to a parent, as I know if their child is hurt and needs hospitalization, it is that child's parent that will be up all night, not me.] I did point out Matthew had [symptoms of] neuroleptic malignant syndrome, and now had tardive dyskinesia, and he was still having head banging with medications. We also pointed out his identical twin developed seizures with meds. One of the staff said only one to two times in the past eight years had they called 911 for a behavior emergency.

Also, during the meeting on campus, the chief compliance officer said Matthew would "probably" attend one of their own day programs as some Opengate residents attended day programs at other agencies. It was uncertain what staff would do if Matthew refused to attend the day program, and they said it had happened with individuals for up to weeks on end. At the day programs, the functioning range was low to high, and classrooms worked on prework skills and did volunteer work. We visited both day programs although our expert had to leave and could not come with us. The program at the residence site had thirty-three individuals, in four classrooms with three staff in each class. We later visited and met with the director of community services at the off-campus day program. At that day program in Hawthorne, there were seventy-four individuals, thirty-four from Opengate, and we were told there was one staff to three people, but it did not appear that way. Two people were fast asleep on the couches, while others were sitting down, with no activity-appearing like zombies, clearly heavily medicated. Someone was walking very unsteadily. In two classrooms people were watching violent movies. I expressed to the director of community services that Matthew had a history of imitating violence on TV programs. We were told individuals were winding down and would not be leaving for forty minutes. The director of community services also stated that staff did not always follow through with the planned activity schedule. She commented on the "male staff not bringing activities to tables" and said male staff were more in tune to "preview behaviors." During the day some individuals go into the community to movies or other recreation activities. [A staff] from JRC stated that if these activities were given to Matthew noncontingently (i.e., not being conditional on desired behaviors), he would have no incentive to behave. Sometimes the day program calls the residence to pick up individuals for behaviors they could not manage, and the director did state that my brother would be "very difficult," "very challenging," and she also stated she wanted to go to JRC to observe Matthew and to advocate for whatever he needed. The director of clinical services at Opengate would not answer our expert's numerous requests to visit the day program.

[Of note, Geraldo Rivera was on the advisory board of Opengate,[26] and with all his public criticism of JRC on his show, broadcast in June 2006, he has said nothing (along with other anti-JRC self-appointed advocates) about the chemical restraint going on.] The staff did ask

26 *Opengate*, Geraldo Rivera 77ABC podcast, January 6, 2012, http://opengateinc.org/2012/01/.

> what Matthew liked, and we told them (zoos). They also said they could not be 100 percent sure they could guarantee his safety and could not be sure he would respond well.

When I visited I made sure to ask very specific, mathematical-type questions that staff could not dance around. For example, the psychiatrist was telling us how she tried to avoid medications and how they could manage one student in particular without medications. When she said she tried to avoid antipsychotics, I asked for the percentage of individuals on antipsychotics. When she told me she was not sure, I asked if it was more than 50 percent. I also asked specifically if there were individuals on three, four, and even up to seven combinations of psychotropic medications.

If a program is run well, the staff is knowledgeable and competent, and my being professional and asking detailed questions should not be intimidating. Staff should understand and respond appropriately to the questions.

One of my first questions was whether, during Opengate's functional behavioral assessments, they used reliable and validated methodology such as Questions about Behavioral Function (QABF) or performed functional experimental analysis.[27] While this was a very important question regarding dangerous behaviors and minimizing or eliminating medication to control behavior, the staff did not even understand the question. I also always ask about making unannounced visits, which are always permitted at the JRC. If a placement is run well, the staff will not care about families making unannounced visits. I also wanted to be friendly to show I was open-minded, so I started by pointing out to the psychiatrist that we worked at the same hospital.

Another agency was also taking individuals back from the JRC. They told the JRC that the residents were doing great and they were seeking to have more JRC residents transferred to their facility. I spoke to the staff from that

27 Functional experimental analysis is a procedure in which an individual is put into different experimental conditions to find out why a problematic behavior is occurring. For example, if the frequency of a behavior increases in a condition when an instructor requires the individual to perform certain activities, the function of the behavior might be to escape from demands.

agency, and I discovered that what that agency called "doing great" involved frequent calls to 911 for former JRC students.

On March 19, 2013, we and the other JRC parents had a group meeting with OPWDD at their request. They reiterated what they had stated in the letter: there was a June 2014 deadline to move all out-of-state individuals back to New York. The parents asked about medications. OPWDD staff initially responded by discussing informed consent for medications.

I took the microphone and explained that if the behaviors were dangerous and the parent did not agree to medication, the hospital and agency could obtain an involuntary court order to administer the medication, and if the parent still did not consent to the medication, the agency and the OPWDD would take steps to remove the parents' legal guardianship. In other words, these people with disabilities would be removed from a place that had been able to manage their behavior without medication that had toxic side effects and would now be placed in programs with inadequate behavior plans, given medications that had potentially lethal side effects, and if parents did not agree, they could lose their legal guardianship. Their attorney, who was present at this meeting, confirmed this.

I saw this happen to another family. An OPWDD-funded agency residence took my cousins to court to remove their legal guardianship from their son with autism after the parents refused to consent to additional psychotropic medication. The judge ruled that if they did not consent within seven days to any medication proposed by a psychiatrist of the agency's choosing, a court-appointed guardian would sign the consent in their place. As of July 2015, my cousins' son takes fourteen medications, most of which are psychotropic or to manage the side effects of the psychotropic medication.

| My Cousins' Son's Forced Medications as of July 2015 | |
|---|---|
| **Medications** | **Treating Diagnosis** |
| Clozaril (Clozapine) 100 mg in the morning and 200 mg at bedtime | Mood disorder not otherwise specified |
| Klonopin 0.5 mg three times daily | Mood disorder not otherwise specified |
| Neurontin (gabapentin) 400 mg three times daily | Mood disorder not otherwise specified |
| Protonix (pantoprazole) 40 mg twice daily with meals | Treatment for gastritis/gastroesophageal reflux disease |
| Depakote (divalproex sodium) 1,000 mg at 5:00 p.m. and 750 mg twice daily | Mood disorder not otherwise specified |
| Senokot 17.2 mg at bedtime (docusate sodium and sennosides) | Prevent constipation |
| Melatonin 6 mg at bedtime | Induce sleep |
| Crestor (rosuvastatin) 5 mg at bedtime | Treat hypercholesterolemia |
| MiraLax (polyethylene glycol) 17 mg at bedtime | Prevent constipation |
| Dulcolax (bisacodyl) 100 mg TID | Prevent constipation |
| Lithium 300 mg in the morning and 600 mg at bedtime | Mood disorder not otherwise specified |
| Synthroid (levothyroxine) 25 mcg once daily | Treat hypothyroidism |
| Vitamin D 50,000 IU once a week | Treat low levels vitamin D |
| Cogentin (benztropine) 0.5 mg twice daily | Control drooling |

I also noted that, with my daughter, when I did not agree with the school district on an appropriate placement, I had the right to an impartial hearing. I wanted to know what rights I would now have for my brother. I was told that because he had turned twenty-one before a certain year, we could have an administrative review (which most parents at present could not have). However, that was the same "kangaroo-court" proceeding we had years ago, with no independent judge but just a panel of other OPWDD employees. This is akin to the judge, jury, and executioner being the same. Fundamental to the American system of governance is due process and a fair and impartial hearing, thus the need for neutrality.

During the meeting, the OPWDD staff put on a PowerPoint presentation about returning individuals back to New York. They mentioned bringing in experts to help us with placement. They reasserted the June 2014 deadline to move all out-of-state individuals back to New York.

I said I had an expert who was willing to contribute, but Opengate was refusing to let her in. I asked what the problem was with allowing her to visit, if the placement was appropriate. I also mentioned my daughter with autism and that I had visited fifteen to twenty different programs recommended by her school district and had never had a problem with bringing in an expert before. OPWDD then contacted Opengate, which finally allowed our expert to visit.

The presentation also mentioned that if a parent did not agree with the proposed placement, funding would be cut in ninety days both for students who turned twenty-one after June 30, 1999, and for students who turned twenty-one before that date, such as my brother. We would only be entitled to the same "kangaroo-court" proceeding we had back in 1995. Therefore, without due process, the state could put someone anywhere and call it appropriate, and the parents would be forced to consent at financial gunpoint.

At the meeting, I also told Dr. Jill Pettinger from OPWDD that it made no sense from a clinical standpoint that PhD board-certified behavior analysts (BCBAs) who were doctors were not allowed to supervise either intrusive or restrictive behavior plans, but the system allowed master's-level social workers to do so. She told me she agreed with me and said the regulations could be amended once BCBAs were able to be licensed. However, BCBAs have been licensed in New York since January 2014, but the OPWDD regulations as of October 2016 have not changed.[28]

I was informed by JRC staff that during another meeting with the JRC, Glenda Crookes, the executive director of the JRC, asked the OPWDD what would happen if it could not locate placements in New York for all the individuals over the age of twenty-one at JRC by June 2014. The answer from Jill

---

28 14 CRR-NY 633.16NY-CRR: Official Compilation of Codes, Rules and Regulations of the State of New York, title 14, Department of Mental Hygiene Chapter XIV, Office for People with Developmental Disabilities, part 633, Protection of Individuals Receiving Services in Facilities Operated and/or Certified by OMRDD, section e, 3i. These regulations include behavior plans that OPWDD considers intrusive or restrictive as defined in section c8, https://govt.westlaw.com/nycrr/Document/Ieb7ffe02978311e29e9f0000845b8d3e?viewType=FullText&originationContext=documenttoc&transitionType=CategoryPageItem&contextData=(sc.Default).

Gentile, then the associate commissioner of the OPWDD, was that they were informed by the governor's office that she and other OPWDD staff would lose their jobs.

When I went to the state capital weeks later, I asked elected officials and aides in the legislature how the OPWDD could make objective decisions whether any placement was appropriate when they had been told they would lose their jobs if they could not locate a placement by June 2014. At that time we could not meet with the real decision makers at the governor's office, only the underlings who had to carry out orders. This completely left the family members out of the decision-making process.

On May 21, 2013, one of the other JRC parents, who had been trying repeatedly to call Opengate to make an appointment after she received a notice for a possible admission for her child, finally got through and was told Opengate was no longer considering JRC students. That was a relief. The parent asked why no one had notified her, and the Opengate staff said it was not Opengate's job to do so. This was correct. It was not the facility's responsibility, but it showed the state was not doing its job in notifying us.

The OPWDD finally did notify us in a letter dated September 23, 2013, written by Donna Limiti, director of OPWDD Region 4. An additional letter from the OPWDD written by Jill Gentile, dated September 26, 2013, was sent to all JRC families and to the state legislators we had asked to hold a public hearing. The letter stated this:

> Currently, your family members are at a school. The school is intended to serve students until they are 21 years old. OPWDD doesn't have the long term authority to fund individuals in school settings. Your children, as all children, should be afforded the opportunity to receive age appropriate services and move on to their adult lives. The services offered by New York State providers are far more varied than those offered in school settings.

This was far from the truth. The JRC-in addition to providing a school-had group homes individually licensed and regulated by the Massachusetts

Department of Developmental Services as adult group homes, and the JRC's adult day program included employment- and community-based day services, also licensed by the Massachusetts Department of Developmental Services. In addition, the JRC had been a licensed adult residential facility for more than thirty years.

In another apparent fabrication, Jill Gentile wrote, "I know many of you hear rumors about individuals returning from JRC with ill consequences but contrary to these rumors, these individuals are living full lives in the community." She made this assertion despite the deaths of former students who had come back to New York, as well as others who had had severe self-injuries, assaultive behaviors, psychiatric hospitalizations, and psychiatric medications that they did not have while at the JRC. The untruthfulness by the OPWDD demonstrates it cannot be trusted to judge if a program is appropriate.

## My Brothers' Fortieth Birthday

On December 6, 2011, I called my brothers on their fortieth birthday. Matthew, at the JRC, told me about the vanilla and chocolate birthday cake he ate and the new electric razor he received as a present. I then called Stuart, who was in St. Vincent's Hospital in Westchester. He told me if the facility did not return a certain recliner, which they had removed because he was compulsively gluing it, he would throw chairs and put a hole in the wall when he returned to his group home.

My feelings that night were so mixed. I was really happy for Matthew but felt very sad for Stuart. He had no quality of life and still does not.

The hospital psychiatrist was annoyed that the group home removed the recliner. I thought it should have been removed because his behavior was inappropriate with it. Due to his threatening behavior, the group home did not want to take him back, but they wanted his medication increased temporarily, even though he was already on high dosages.

The hospital discharged him back to the group home although he was making threats. Their justification was there was nothing they could do for him. Days after his discharge, without our knowledge, the group home staff returned the recliner to him. My mother found out because Stuart called her one night stating he did not go to his day program but was messing up the recliner instead.

We can take Matthew out to restaurants and museums. One time I was deeply touched at a restaurant. When my mother went to pay the bill, the waitress told us that the people sitting at the next table had a relative like Matthew and they paid our bill. We have gone to Vermont with Matthew, and he has attended my wedding, our grandmother's one-hundredth birthday party, and my daughter Batsheva's bat mitzvah.

Meanwhile, his identical twin brother is heavily medicated and has life-threatening side effects from his medications. Due to his life-threatening aggression and self-injury, he has been unable to attend some family celebrations, such as our grandmother's birthday party, even though it was near his group home. These are examples of how dangerous behavior results in a more restrictive environment and precludes community integration.

While Matthew is awake and alert during the day, doing vocational work and going on trips with his family, Stuart is sedated, sleeping much of the day, angry and unhappy with his home environment. He has repeatedly stated he wants to live with his brother in Massachusetts. While I can enjoy taking Matthew to museums, zoos, and restaurants, to take Stuart anywhere could be dangerous.

If the government can take away one research-validated effective treatment-even if it is the only effective treatment-from one person with one medical condition, then the government can take away any effective treatment from anyone for any medical condition. There is such a political stigma against the JRC because of its use of aversives that others will not even study JRC's systems or try to implement their positive-behavioral interventions. For example, in JRC classrooms with low-functioning students, the computers are attached to vending machines to provide frequent reinforcement for students who need it without necessitating staff intervention. This kind of technique could be provided in other settings where limited staff is available.

One time when I was driving with Matthew, I put on some music, and he started rocking back and forth. I told him, "No rocking." He corrected me, telling me, "Rock to music." At the JRC, the staff realizes it is important to allow people some time to engage in stereotypy (mechanical repetition of speech or movement), as long as it is not dangerous and does not interfere with learning. Therefore, Matthew receives opportunities to listen to music, and he can "rock to music."

In conclusion, Matthew's life was saved because repeated head hitting against sharp objects, with resulting open, bloody, head wounds and trauma, if continued, could have killed him. He was also saved from a life of restraint, mind-numbing drugs and hospitalization. Intensive positive reinforcement supplemented with rare use of skin shocks not only saved his life but also let him function and give him quality of life. Outside efforts to stop his treatment have been unsuccessful. However, his identical twin brother could not obtain access to effective treatment.

# Three

## Stuart Lost His Quality of Life

This chapter focuses on Stuart, how he once did well and even exhibited some level of independence in a structured school setting and later at a full-time job, but in later life, a lack of structure sent him on a downward spiral even putting his life in danger.

### Behavior Therapy with Rewards and Consequences Fosters Independence and Inclusion

I remember Stuart as playful, even though it was on his terms. He once made up a self-stimulatory game with repetitive vocal utterances and hand movements and asked me if I would like to participate because he wanted to take turns. He would ask me to do things like wash windows with him. I used to take him for rides on the subway and go places. We played Uno.

I once baked oatmeal-raisin cookies for a party at my daughter's school. It brought back memories of baking the same kind of cookies with Stuart. I remember suggesting to him that we bake cookies, and he chose oatmeal-raisin cookies. We worked well together, and our cookies came out great.

He was teased by the children on the block when he was young. One day when he was inside, I spoke to the children and told them nicely about his condition, that he had a problem with his brain and that it was not his fault he would say the things he did. The children asked me some questions about it, and they never teased him again. Stuart attended the Clearview School in Westchester after the Reece School could not manage his behavior. At Clearview, he had good behavior structure, and he was generally well behaved. He did rip up favorite books, but he was not violent to anyone, including himself.

It was difficult to get funding for Stuart for the Clearview School, especially after an IEP meeting when school was about to start. My mother never received the final notice of recommendation. She called the COH to find out where it was, and she was told it was in one of the employee's drawers.

After graduating at age twenty-one, he began to have problems. He attended a day program at an organization called AHRC (the Association for the Help of Retarded Children), which started as part of another organization called ARC (Association for Retarded Citizens) and had a history of being against aversive treatment. Stuart went to the program by himself by bus. However, there were problems there. He sometimes had nothing to do and once told us he was spending time watching trains outside, where he was unsupervised. Behavior problems at home began to set in. He started to become violent. He had psychiatric hospitalizations and medication management. He then attended the day-treatment program at a local hospital.

Psychiatric hospitals and day programs carry risks to individuals with intellectual disabilities. When he attended the psychiatric hospital-based continuing day-treatment program, he started to pick up ideas from general psychiatric patients, started talking about stabbing himself as another patient had done.

One day the staff called my mother to tell her Stuart was hiding in the bathroom and they would have to admit him into the hospital. My mother informed the staff that when my brother was having an asthma exacerbation, he engaged in this behavior. The staff insisted that my brother had no problems with his asthma. My mother drove to the hospital with his asthma medications and found him red all over and itching. She gave him his inhaler, and he was fine. He did not have to be admitted to the psychiatric hospital.

He later had a full-time job at Meal Mart, where he swept floors all day. He told me there was always a mess to clean up. This job lasted two and a half years.

This job was one of the best things to ever happen to him. The structure kept him safe. He was off all medications. He was going out to eat and to the movie theater with two special-needs friends on the weekends without supervision. He would stay at my house when my parents were on vacation. He would make his own lunch the day before, get up at six in the morning-while my husband and I were still asleep-get dressed, make Toasty-Cakes for breakfast, take his lunch from the refrigerator, and then take three public city buses to get to work, all on his own. I told him he needed to leave by six forty to get to work on time, and he always left before I even woke up.

Once during the winter, my husband offered to drive him to work, but Stuart wanted to take the buses. There was only one instance of trouble, which is when he told me on his way home that a driver shut the door in his face and would not let him in, so he simply took the next bus. He was expected to and would help me cook dinner in the evening and help with chores on the weekend.

After I had my first daughter, Batsheva, he stayed with me while my parents were on vacation. He was helpful and would hold my daughter on his lap while I cooked dinner. He would watch her in the living room while I was busy in another part of the house and would call me if she started to cry.

My husband or I would take him to watch some subway trains on Sundays as a reward for appropriate behavior. However, if he had even a mild behavior problem, there were clear and immediate consequences. For example, he once had to use a particular bathroom at our synagogue, which was occupied. He angrily banged on the door, and for that he lost access to his music and subway books for a day. He stated he was sorry for what he did, and he never did it again.

He responded very well to limit settings, clear expectations, and responsibilities, which were more than my parents gave him. Once when my parents returned from a vacation, I told him to get his Toasty-Cakes to take back home. While retrieving them from the refrigerator, he said, "Sorry, Toasty-Cakes, you have to go back to the Bronx." During the time he was employed, he moved into a New York State OPWDD-funded residence. Initially, he was a model resident. Problems began after a coworker teased him, repeatedly stating, "You have to work until five today." My brother had a fixed routine of leaving his job at 4:00 p.m. Stuart became very agitated and ultimately grabbed the coworker's butcher knife. My brother was fired.

## How "Positive" Behavior Supports Destroyed Stuart

With his job loss, Stuart lost his structure, and the group home had no mechanism to replace it. In New York State, a group home can refer a resident to a psychiatrist or even a nurse practitioner for medication, but his train books or music could not be taken away to modify a behavior because that was "his right." Removing preferred items is in conflict with the "positive" behavior support movement, which defines an aversive as anything that causes physical or psychological pain. According to the supporters of this theory, aversives should never be used.[29] Although in real life, people are repeatedly told "no," that they cannot have something they want.

After Stuart's "positive" behavior treatment, his life was destroyed by dangerous behaviors, repeated hospitalizations, psychiatric medications and more medications due to the adverse side effects. My brother became obsessed with fires, and once tried to set himself and a peer's hand on fire. He craves limits.

---

29 "The current revision of the TASH resolution (*TASH [resolution opposing the use of aversive and restrictive procedures*], 2002) continues the conflation of 'aversive' with 'abusive,' and…'the definition of 'aversive procedures' is expanded by the addition of two items, 'rebellion on the part of the individual' and 'permanent or temporary psychological or emotional harm,' items that invite arbitrary decisions about whether or not a certain procedure is aversive." Crighton Newsom and Kimberly A. Kroeger, "Nonaversive Treatment," in *Controversial Therapies for Developmental Disabilities: Fad, Fashion, and Science in Professional Practice*, ed. John W. Jacobson, Richard M. Foxx, and James A. Mulick (Mahwah, NJ: Lawrence Erlbaum Associates, 2005), 408.

Sometimes I have called him, and he has refused to get on the phone, yelling in the background, "I'm on LOP." I then would find out that earlier he had exhibited a dangerous behavior. Stuart apparently had learned this from his brother, because at JRC, LOP has always stood for "loss of privileges," and phone calls have to be earned.

Stuart did not fail the limited positive-behavior supports available; the limited available positive-behavior supports failed him. Sometimes he has refused attending the day program completely or insisted on going in late. The current "positive" behavior plan has not been able to get him to attend consistently, and the lack of structure results in dangerous behavior. He was even positively reinforced for attacking a peer at day program by being suspended for a week. (Not having to attend the day program during his suspension was a *desirable* consequence from his point of view.) We cannot get the help he needs thanks to the anti-aversive zealots. He is over twenty-one and therefore has no legal right to an appropriate placement, utilizing effective behavioral intervention.

Stuart has been taken away from the residence in handcuffs by police. However, this was also inadvertent positive reinforcement because he loved the attention. Sometimes 1:1 staffing has only served to increase the frequency of his dangerous behavior, because he likes to ask provocative and threatening questions, seeking the listener's reaction. He knows he can get 1:1 or even more staff by exhibiting dangerous behaviors.

His first inpatient hospitalization at the residence was soon after he was fired from his job. The psychiatrist, without speaking with the family or residence staff, put him on a cocktail of Tenex (guanfacine), Risperdal (risperidone), and Paxil (paroxetine). I called her and stated that the family needed to give consent and we had not. She retorted that he was an adult and could make his own decisions. If she had properly assessed him, she would know that there was no way he could understand the risks or benefits of medication treatment. When I told her to contact the previous treating psychiatrist, she told me to have the treating psychiatrist call her. I then contacted the hospital's risk management department and faxed over the legal guardianship papers.

My husband left a voice mail with the psychiatrist and told her that he was a lawyer and that the family had a legal guardianship and needed to give consent. He then suggested she contact her own lawyer before calling him back.

She resigned from the case instead, and a new psychiatrist took over who worked more closely with us. However, he only did inpatient-not outpatient-cases, so we took Stuart back to his former psychiatrist, even though her office is far away.

When outside of the residence, Stuart has behaved violently with his family. Once at a family gathering, Stuart became agitated, obsessed with taking hanging threads out of clothing, and hit our grandmother in the face. My parents have tried to bring him to Massachusetts to visit his twin, but even this has been too dangerous. Once in a restaurant, Stuart wanted chocolate pudding, but because of his obesity from the psychotropic medication, my mother refused to order it. My brother responded by placing his hands around her neck.

One time while my parents were out of the country, Stuart again exhibited dangerous aggression at the residence. The staff was considering increasing his medication and even hospitalization. I had to decide if I should contact my parents or not. I felt I had no one to look to for help in this decision. However, I decided not to contact them. I thought they should enjoy their vacation, and there was nothing they could do anyway. Instead, I consented to increasing his Risperdal, believing I had no other choice. Ultimately, Stuart was not hospitalized at that time, but Risperdal, like other medications, later proved ineffective.

During one hospitalization, his hospital clothes appeared dirty. When I asked when he last changed them, he replied it was a week before. I told the nurse that he needed reminders to change his clothes regularly.

A five-month hospitalization stay was precipitated by something as simple as a staple being out of place on a chair. He initially appeared calm when reassured, but fifteen minutes later, he unpredictably left the residence, ran into the street, and took a swing at a stranger. The state refused to place him at the

JRC with his twin brother despite that one of their own psychologists, sent from the Institute for Basic Research, evaluated Stuart during that hospitalization and subsequently told my mother that if anyone had a case for the JRC it was Stuart. I asked him to put in writing, but the psychologist said he was not sure if he was allowed to do so. The psychologist never wrote that recommendation. Instead the state wanted to place him at the Brooklyn Developmental Center.

When I went to visit the Brooklyn Developmental Center, I saw a few residents in the lobby, clearly overmedicated and too unsteady even to stand up. One of the residents sat down in a chair and fell asleep with her head leaning back into a glass cabinet. Someone who appeared to be a staff member was slapping her in the face and yelling at her to get up. When she would not wake up, the staff person just left the area. I do not know who the person was because no one wore identification badges. The security guard walked away from the desk, so anyone could walk in or out.

In the sleeping quarters, there was a resident handling cleaning fluid without gloves. When the staff person with the resident saw us, she left to get some gloves. My aunt noticed the label on the bottle and saw it was a toxic substance. When she inquired further, we were told that there was a different fluid in the bottle than what was on the label. The residents' bathroom had a foul smell.

I was told that the direct-care staff did not even need a high school diploma. When visiting one of the workshops, a resident approached me in a threatening manner, and we were told to leave immediately. In that brief visit, I saw that many of the residents had nothing to do. Overmedicating to the point that a person can barely stand up or falls asleep during the day should never happen. Furthermore, the absence of identification badges in a large facility means there in no accountability for mistreatment.

For these reasons, my family disputed with the OMRDD over Stuart's placement. If there had been an appropriate placement to begin with, Stuart would not have lived in a hospital for so long. After my parents paid thousands of dollars to an attorney, the OMRDD finally agreed to return Stuart to his group home. Temporally he received 1:1 staffing.

Although the group home is limited in the behavior plans it can implement, the staff are kind and have gone out of their way to help. Staff members have taken him on individual trips to different parts of the country, including to Disney World in January; these trips were designed to decrease the obsession Stuart had with Christmas lights going away at the end of December. This obsession has caused him to engage in dangerous behaviors.

However, Stuart has been on at least sixteen psychotropic medications or medications used for behavioral reasons that have been FDA approved only for nonpsychiatric medical reasons. These medications include Celexa (citalopram), Zyprexa (olanzapine), Paxil (paroxetine), Seroquel (quetiapine), Depakote (divalproex sodium), Lamictal (lamotrigine), Geodon (ziprasidone), Thorazine (chlorpromazine), Risperdal (risperidone), Abilify (aripiprazole), Inderal (propranolol), Effexor (venlafaxine), Catapres (clonidine), Tenex (guanfacine), Ativan (lorazepam), and Desyrel (trazodone). As I noted earlier, Stuart has tardive dyskinesia from his long use of neuroleptics. He was placed on Dilantin (phenytoin) when he had seizures after being put on psychotropic medications; he had never needed these in a structured school setting or when he was gainfully employed. He developed obesity and sedation, sometimes sleeping until midafternoon. It is painful to see him with all these side effects.

As of August 2015, he is on twenty regular medications plus five as-needed medications. His medications are listed below:

| Stuart's Medication List as of August 2015 |
|---|
| Lamictal 150 mg twice daily |
| Catapres 0.1 mg at bedtime |
| Thorazine 150 mg twice daily |
| Seroquel 300 mg twice daily |
| Desyrel 50 mg at bedtime |
| Effexor XR 150 mg at bedtime |
| Amitiza (lubiprostone) 24 mcg twice daily |
| MiraLax powder 17 g twice daily |
| Vitamin C 500 mg once daily |
| Kristalose powder (lactulose) 17 g twice daily |
| Mineral oil two tablespoons three times a day |
| Senna 34.4 mg at bedtime |
| Colace (docusate sodium) 400 mg at bedtime |
| Philips Colon Health Probiotic one tablet daily |
| Omeprazole 20 mg daily |
| Bisacodyl 10 mg twice daily |

| |
|---|
| Vitamin D3 1,000 IU once daily |
| Multivitamin two tablets at bedtime |
| Advair Diskus (fluticasone propionate/ salmeterol xinafoate)100/50 mcg twice daily |
| Singulair (montelukast) 10 mg at bedtime |
| Benadryl (diphenhydramine) 25 mg as needed for allergy |
| Bentyl (dicyclomine) 20 mg one tablet twice daily as needed for gastrointestinal upset |
| Simethicone 80 mg three times daily as needed for gastrointestinal upset |
| Claritin (loratadine) 10 mg as needed daily for allergy |
| Patanol opthalmic (olopatadine) 0.1% solution eye drops twice daily as needed for allergy |

For 2015, the total price of all of his medications was $21,585.32. All six of the medications to manage his behavior have constipation as a side effect. He is therefore on multiple types of laxatives, including stimulant laxatives, which have their own long-term side effects, such as weakened colon muscle tone (which can become permanent) and osteomalacia (softening of the bones), vitamin and mineral deficiencies, and fluid and electrolyte imbalances. April 2016, Stuart had to go to the emergency room for constipation despite the use of all these laxatives. Most of these medications are either psychotropics or to treat side effects of other medications. Despite all these medications, his current behavior plan does not even have a functional behavior assessment, a procedure done to try to find the purpose of the behaviors, because the residence has informed me OPWDD only requires them if the individual needs physical restraints. My brothers are very close. Because of their limitations, they need an adult to bring them to visit each other. It is wrong to separate them, but because of the anti-aversive movement, we cannot find a group home for Matthew in New York to keep him safe, and the OPWDD refuses to place Stuart in Massachusetts. I find it particularly sad because, as it is, they have social limitations and cannot travel independently because of their autism and intellectual disabilities.

In summary, for Stuart's dangerous behaviors, medication has replaced education, and he lost his functioning and quality of life. Although he was once higher functioning and more independent than Matthew, the lack of effective treatment has made him more impaired than his identical twin. He has gone from a minimally restricted environment, commuting to work by himself, and going out with friends to requiring constant supervision, which is very restrictive. So much for *least restrictive environment.* Furthermore, in addition to the drugs' limited effectiveness, the side effects of the drugs have given him a multiple medical problems and even endangered his life.

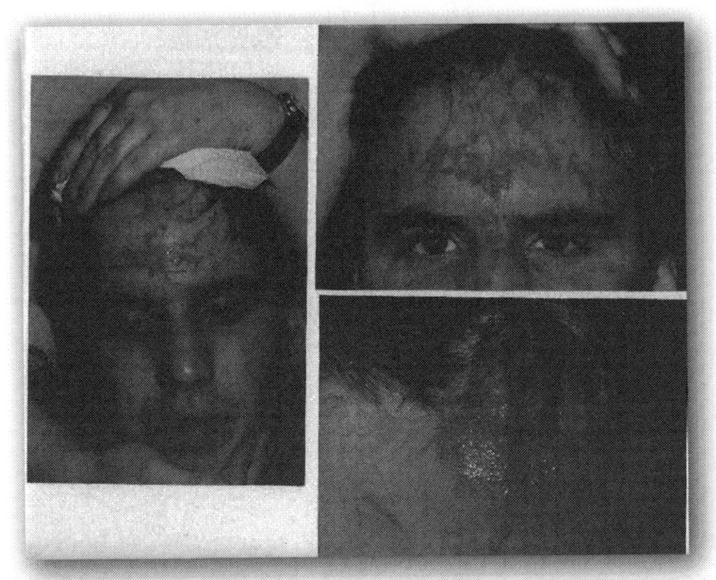

My brother Matthew before aversive therapy.

Matthew after aversive therapy.

**Matthew, his twin Stuart, and my mother, Lorraine Slaff. Matthew won a medal at the Special Olympics.**

**My husband, Albert, and I after completing the Cape Cod Marathon, sitting with Matthew. We raised $4,685 for the JRC.**

# Four

## Talia-Autism All Over Again

Talia diagnosis was a shock, and it was a long process to learn what effective evidence-based treatment was and how to obtain it. I will be discussing effective treatments, how I obtained them, and the results from my efforts, progress that most children make without intensive treatment and that most of us take for granted.

### Early Signs of Autism

For years before I was ever married, let alone pregnant, I had nightmares of having a child with autism. I thought I could never do what my mother did to raise my brothers. I thought I was not as strong as her. My husband and I had a genetic workup, and we were told our chance of having an autistic child was only 3 percent. Finding out on a pregnancy ultrasound that we were having a second daughter was reassuring. Before our daughter Talia was born, everyone in the family with autism was male. Although I would never ever do it, the thought of aborting a pregnancy if the child was a son did cross my mind.

When Talia was not even a year old, my husband noticed she spun her toys a lot. He thought this was a sign of fine motor coordination. However, my first thought was autism. I did not share that thought with him. I thought I was just worrying too much because Matthew and Stuart were autistic.

When Talia was a year old, I could get lots of things done around the house. It was different than with our older child, who would cry when she was not held. I discussed this with my mother and an aunt. They both reassured me that every child was different and that my Talia was not handicapped. It made me feel better.

Now when I look back, I realize that Talia never looked at me when I nursed her.

At about twelve months, she waved bye-bye once, but never again. At fourteen months, she appeared to call me Mama twice, but never again. I became suspicious at fifteen months when she had not said a word.

At the fifteen-month doctor visit, I told my pediatrician I was concerned about her lack of language. She reassured me, told me I could wait another six months, and to see what would happen. This time I would not listen.

## Initial Home-Based Therapy

I called an early intervention agency. The psychologist who came suspected Talia had pervasive developmental disorder-a general term for a group of disorders defined by autistic features. He said we could start early intervention with special education for three months and then reevaluate.

If it was an autism-spectrum disorder, we would start applied behavioral analysis, and he could diagnose an autism spectrum disorder immediately. He said this was a borderline case. I initially told him to just do standard special education for now.

I then called my mother, who insisted I call the psychologist and tell him to diagnose her with autism right away. She said, "Who cares when she is an A student in college if she had an autism diagnosis at the age of sixteen months?" This was excellent advice. It turns out my daughter has severe autism.

My husband had been diagnosed with diabetes a few weeks earlier. After these diagnoses, I would be forever like a firefighter-it always seems I am putting out a fire. At first, I cried a lot. It felt like the child I thought I was having-the child I thought I had-had died.

She was diagnosed early in the morning. I had to be at work at eleven o'clock for my job as a psychiatrist in an emergency room. I cried while driving to work, but when I was close to the hospital, I knew it was time to stop. I remember going into my supervisor's office without knocking to ask a question. I immediately apologized and later told him what happened. He was really good about it. He asked me if I wanted to go home, but I told him that listening to my patients made me forget about my daughter, which was true.

After her initial diagnosis, our life became a series of doctors' appointments, therapies, hopes for progress, disappointments, and many other things. I did not even know where to start.

When I was a child, I always wondered when my brothers would talk. For example, I remember thinking to myself when I was four that I talked, so they should talk when they were four. I wondered and prayed that my own daughter would talk one day.

While New York City attorneys might not recommend that they be present at early intervention meetings, when I went to my meeting, I took an attorney with me, even though she did not specialize in special-education law.

We immediately received twenty hours of applied behavior analysis (ABA), with an increase to twenty-five hours in one month and to thirty hours two months after that. Eventually we had thirty-four hours.

I have met parents who have told me their child received only ten hours of ABA, despite New York's own early intervention guidelines, which suggest a minimum of twenty hours of ABA.[30] The National Institutes of Health (NIH)

30 New York State Department of Health, *Report of the Recommendations - Autism / Pervasive Developmental Disorders, Chapter IV (Continued) - Behavioral and Educational Approaches, Intensive Behavioral and Educational Intervention Programs, Recommendations*, accessed September 4, 2016, http://www.health.ny.gov/community/infants_children/early_intervention/disorders/autism/ch4_pt2.htm.

recommends a minimum of twenty-five hours a week of intervention.[31] The fewer hours of service should not be allowed. Providing fewer hours is not cost effective, because more of these children will need long-term special education and later require residential treatment.

Talia has been through years of Lovaas ABA, a method that utilizes *discrete trial training*, which breaks a skill down into multiple small steps.[32] Discrete trail training teaches one step at a time with repetition while gradually reducing prompting, until the step is mastered independently, using reinforcers after correct responses for motivation or a negative consequence such as "no," if the response is incorrect.[33] Lovaas ABA also utilizes *incidental teaching*, teaching skills in the natural setting where and when they usually occur to promote generalization.[34] Data are collected.[35] However, Talia did not always have the best therapists. When I look back, I see I should have hired an outside consultant when her progress was minimal. It is a travesty that there is no minimum training or experience requirement to provide ABA services. The agency I used for Talia never used a board-certified behavior analyst (BCBA) or closely monitored what was going on.

---

31 National Institute of Mental Health (NIH), "A Parent's Guide to Autism Spectrum Disorder," US Department of Health & Human Services, National Institutes of Health, publication no. 11-5511, revised 2011, http://www.nimh.nih.gov/health/publications/a-parents-guide-to-autism-spectrum-disorder/index.shtml.; Lovaas recommended "[u]p to 40 hours per week for as much as 2 years may be needed for optimal gains," O. Ivar Lovaas, "Behavioral Treatment and Normal Educational and Intellectual Functioning in Young Autistic Children," *Journal of Consulting and Clinical Psychology* 55, no. 1 (1987): 3-9, and "Todd Risley...found the rate of verbal interaction that Lovaas estimated to occur in 40 hours of Early Intensive Behavior Intervention (EIBI) per week to correlate well with the natural rates of interaction that Hart and Risley (Betty Hart, Todd Risley, *Meaningful Differences in Everyday Parenting and Intellectual Development in Young American Children*, (Baltimore, Maryland: Brookes, 1995) had observed in very young typical children...Risley surmised that the modest gains made when EIBI is implemented for fewer than 40 hours per week might result from the lack of a critical number of adult-child interactions." Betty Fry Williams and Randy Lee Williams, *Effective Programs for Treating Autism Spectrum Disorder: Applied Behavior Analysis Models* (New York, NY: Routledge, 2010), 238-39.

32 Fry Williams and Lee Williams, *Effective Programs for Treating Autism Spectrum Disorder: Applied Behavior Analysis Models*, 111.

33 Ibid., 111-12.

34 Ibid., 107.

35 Ibid., 116.

Later on, an incompetent BCBA supervised Talia's programs. She would unexpectedly not show for sessions. The quality of her work was poor. We had to be trained and knowledgeable to know what makes an effective therapist or supervisor and how to listen to others who question the quality of other people's work.

## Obtaining Effective Preschool Special-Education Services

Once a child turns three years old, the funding sources change from early intervention to the school district. I knew I needed a special-education attorney in case the CSE recommendation was inappropriate. My mother suggested I call my brothers' old school, the Clearview School, and talk to the director. The director told me he would recommend someone. I got off the phone and was in tears. All those memories of my brothers came alive again, and here we were going through the same problems. The director contacted an attorney who formerly worked for Clearview School. The attorney was very nice but now worked for a school district, and she told me she could not represent me for that reason. But she was able to tell me who to use and not to use. She told me that the Clearview director had told her to keep a special eye out for me. She called the person who would become my lawyer just to make sure someone responsible was handling this case.

I wanted Talia to have a private evaluation to determine what she needed. Furthermore, some private evaluators will testify on your behalf at impartial hearings or higher courts if necessary. I knew if I let a Department of Education-funded agency do evaluations, they would likely recommend their own placement, whether the placement was appropriate or not. I did not realize that there was a three-month wait for an appointment for a private evaluation, and my IEP meeting was less than three months away. I used the McCarton Center in New York and was lucky they suddenly had a cancellation before my IEP meeting. It was costly, several few thousand dollars, but the results were well worth it. The McCarton Center recommended 1:1 ABA.

However, it concerns me that appropriate education has become boutique care, where usually only cash-paid physicians will testify at hearings as medical insurance will not pay for this service. At the Medicaid clinic where I work, I once needed to testify at an impartial hearing for a patient whose current placement could not address this individual's behavioral needs. I requested half an hour in the schedule to testify by phone. The practice manager told me I could not testify that way because Medicaid would not pay for it, so I had to get permission from the chief of psychiatry.

In 2005, at age three, Talia also needed a social component to her programming. First, it would be good for her to learn appropriate behaviors from typical children. Second, the school district could argue that a completely home-based program would lack a social component, and they could therefore refuse to grant it. I got her into a preschool a few hours a week with a therapist shadowing her. I was nervous during the preschool testing, and I explained to the staff that Talia would have either me or a special-education teacher with her at all times. They were very accepting of Talia.

When I went to my first IEP meeting and told the administrator that Talia had undergone private evaluations, she quickly permitted our continuing the home-based program for thirty-five hours a week-although I began by asking for more hours. Wishing a school would meet her needs, I visited schools that my administrator told me to see prior to our meeting. At these visits, I brought an ABA therapist with me, had a written list of questions, and documented very carefully the answers and my own observations. I was therefore able to demonstrate at the meeting why each preschool program was not appropriate for Talia's needs and could do so if necessary in court. To prevail in court, a parent must be able to provide such reasons.

Although I was prepared for my IEP meeting, I had an "IEP meeting in reverse." Unlike most IEP meetings, in this instance, the district administrator agreed to the recommendations, and the IEP was completed before the administrator asked if she could have a copy of Talia's evaluations. I had mailed those evaluations to her weeks before, and I reassured her they were in my daughter's file.

For Talia's second year with the Department of Education (DOE), I continued a full home program, increased Talia's ABA services to forty-five hours a week (including supervisory hours and additional team meetings), and requested a 365-day IEP. I do not understand why severely autistic children need four weeks off in the summer, and one and a half weeks for winter and spring breaks. My daughter does not even know what a day is.

What her therapists and I know is that when Talia misses a week of therapy, she regresses by one to two months. I have letters from the therapists stating this, but the DOE does not appear to care and is too rigid in its schedule. The DOE administrator told me such services did not exist. In addition, successful ABA in completely home-based case for severe autism requires two-hour team meetings twice a month. The DOE administrator told me there was no funding for team meetings.

I had my attorney request an impartial hearing. My attorney asked the DOE for an IEP. The DOE sent her one, but I had never seen or heard of one like this. The area where it said "date of meeting notice sent to parent" was left blank. This was a clear violation of New York State law, which stated a parent or legal guardian must have five days' notice prior to an IEP meeting. Furthermore, the IEP only had speech goals, but no cognitive, toilet training, occupational therapy, or physical therapy goals.

I brought my ABA therapist to the impartial hearing and had Talia's speech therapist and developmental pediatrician available to testify by phone. They never had to testify. After my attorney showed the DOE's attorney the IEP and asked her how she could defend this, their attorney recommended full settlement.

By law, parents can hire their own providers, pay privately, and seek reimbursement, but they must demonstrate the need. Reimbursement takes months or years. For Talia, it has been more complicated. I started with one private therapist for seven hours a week, and then used a second private therapist for additional hours because one DOE-funded agency therapist left suddenly and the DOE did not replace her. Ultimately, I had to pay therapists and wait more than a year for reimbursement.

I have been able to show that a DOE-approved agency therapist was grossly inadequate and that I contacted some DOE-approved agencies, but no one was available. Neither the DOE nor its farmed-out private agency supervised the therapists who came to my home to make sure they were effective.

One of the problems in New York City is that when Talia's source of funding switched from early intervention to the school district, the hourly fee per therapist decreased by twenty to forty dollars per hour. We had some good therapists and did not want to lose their services.

Another complication was that two of my therapists were DOE schoolteachers. Agency consultants provided the home-based ABA therapy; it was considered a conflict of interest to do both. At my first IEP meeting, I was told this was not allowed.

Well, I would not take no for an answer with my daughter's life. My husband and I wrote letters to every politician we thought could possibly help. After writing more than twenty letters, I received about three responses.

One politician's aide gave me the number of someone at an autism organization. When I called the person there, he gave me a name and phone number to call at the DOE. When I called that number, I was given another number for another person, an official at the central office for preschool special education. That person then told me that my district would have to send a letter to her informing her there were no other providers, and she would have to approve it and then send it to the DOE's Conflicts of Interest and Ethics Committee for final approval.

I contacted my district, which had not yet found a therapist. I had to explain to my local Committee on Special Education administrator the waiver process. She asked me to write a letter explaining all the agencies we had contacted and that we could not find adequate providers. I did this. She then wrote a new letter, received approval from the district chairperson, and sent it to the central official.

I was calling my local administrator daily, and the central official kept requesting more information from the local administrator. This was blocking the approval process and, if not taken care of by the time my daughter aged out of early intervention, would leave my daughter without services. After the central official administrator had all the information she needed, she forwarded the request to the Conflicts of Interest and Ethics Committee, but the official there needed even more information.

When I tried to tell him the information he requested, he insisted it had to be through the district and not from me. I then called the district to give the administrator the information. She in turn gave it to the Conflicts of Interest and Ethics Committee. We finally got a waiver.

Later, after we showed the DOE could not provide a qualified therapist, I had my lawyer file an impartial hearing for reimbursement for the private ABA therapists. Among treatments for autism, which is a biological disorder, ABA has received by far the most research and has the most supporting evidence. Imagine someone had cancer, had to lay out tens of thousands of dollars for chemotherapy, and then had to hire a lawyer and sue for reimbursement. This should not be acceptable for autism or other developmental disabilities.

Another problem arose at about this same time. As I noted earlier, when Talia turned three, the funding source for her therapies changed. This means we had to change our speech, occupational, and physical therapists. It was sad and frightening to see them leave. I did not know what was coming next.

When the child is under the age of three and is in the early intervention program, a service coordinator finds therapists for the child. When the child is older than three, parents need to find their own therapists for any services they need outside of school. The DOE provides a long provider list.

## Speech, Occupational, and Physical Therapy That Works for Talia

For speech therapy, I was advised by my developmental pediatrician to get someone trained in the Prompts for Restructuring Oral Muscular Phonetic

Targets (PROMPT) method,[36] because of Talia's verbal apraxia, a motor speech disorder in which the brain cannot accurately plan and coordinate movements needed for expressive language. I found one place in Queens, which was a thirty-minute drive each way, but I was told people even went there from another borough. However, there was a waiting list.

I remember a day in early September 2005, when I was told I would be called and informed if Talia was accepted. I had my cell phone close to me all day. I felt more anxious than when waiting for my psychiatry board exam results. Every time the phone rang, a feeling of terror overtook me. I finally got the call, and she ultimately got accepted.

For the next two years, someone drove my daughter to a center for speech and occupational therapy. However, I learned what a good speech therapist could be with her new speech therapists. I realized that, in the early intervention program, many of the methods the speech therapists used were not effective because they tried to introduce language through play, not through an organized structured system of learning, which was something Talia needed. She also benefited from going to a sensory gym for her occupational therapy.

However, things became tricky. Unlike in early intervention, which was done at home, her new therapy at the center meant someone had to bring Talia to the center, and someone else had to be available sometimes after school for my older daughter Batsheva. Batsheva's preschool agreed to hold her an extra few minutes on Fridays to give me time to return from the center. When my children were both school age, either my mother or I took Talia to the sensory gym after school, and I had the school bus bring Batsheva to a friend's house, and then one of us would pick her up on the way home.

I was always able to find a physical therapist to come to the house. When Talia was older, all the therapists came to the house, which was better for her. But I always had to find my own therapists, either by spending hours on the

36 "PROMPT is an acronym for Prompts for Restructuring Oral Muscular Phonetic Targets. The technique is a tactile-kinesthetic approach that uses touch cues to a patient's articulators (jaw, tongue, lips) to manually guide them through a targeted word, phrase or sentence. The technique develops motor control and the development of proper oral muscular movements, while eliminating unnecessary muscle movements, such as jaw sliding and inadequate lip rounding." PROMPT Institute, accessed January 10, 2016, http://www.promptinstitute.com/?page=FamiliesWIP.

phone or by sending tons of e-mails. Of course, it did not help that early intervention speech, occupational, and physical therapists were reimbursed much more than the DOE paid for half-hour sessions. Although it was stressful and challenging to find my own therapists, my daughter was better off. In the end, the DOE's failure to provide adequate service coordination worked out better for Talia.

Nevertheless, there were complications. I once had a speech therapist announce in March that she wanted to cut her hours in two weeks. She called to tell me while my husband and I were on our way to a religious service, and after I had fasted all day. I was very upset, and my mind was not clear from the fast. I was pleading with her on the phone to stay until the end of the school year when I could more easily find a replacement. While speaking with her, I was crossing a major street, and my husband yelled because I was not watching the traffic.

After the prayer service, when I was allowed to break my fast, I had no appetite. I was consumed thinking about how to find a competent replacement in the middle of the school year. I did not sleep well either. The therapist finally relented, did the right thing, and did all her sessions for the rest of the school year.

That summer I had another speech therapist who ended the sessions early because Talia was attacking her. I was concerned she was providing negative reinforcement for Talia's aggression. This would allow Talia to avoid or escape work by being aggressive, and Talia would learn that aggression could get her out of work. This therapist did not believe in edible reinforcers, although all the other therapists were using them. She then told me that Talia did not tolerate one-hour speech therapy sessions.

My aunt bought me tickets to *The Lion King* on Broadway for that same day. During the play I worried about this therapist. I did not want her to write a report or testify that Talia did not tolerate one-hour speech therapy sessions. I later told the therapist that, for the past three years, Talia had four different speech therapists who provided one-hour sessions and that she was the first one to tell me this. I replaced this therapist.

Two years later, another speech therapist, suddenly had to go to the hospital for an undetermined length of time. Now with e-mail, it was easier to locate a provider. I simply cut and pasted the e-mail addresses from the entire New York City DOE provider list for Queens and only had to call a few therapists without working e-mails. One ABA therapist had a colleague who could work two days, and through all the e-mails, I got a therapist for the other three days. Both of them were PROMPT trained and had autism experience. My brother Stuart remembered his speech therapist's phone number from almost twenty years before and suggested I call her. Because of the constant searching, my daughter went only one and a half weeks without speech therapy.

We had other problems with occupational therapy. An occupational therapist announced in October she was moving in a few weeks. I was disappointed about her unprofessionalism and felt that if she was contemplating moving, she should not have agreed to take on the case in September, as, again, it would be difficult to locate a therapist who could fit into Talia's schedule.

Fortunately, rules changed in New York City, and DOE employees were now allowed to provide services after school as related service providers. I used a DOE provider list for employees of public schools and called some in my neighborhood. This way I was able to obtain an occupational therapist to come to the house to work on Talia's activities of daily living (ADLs) as well as fine motor skills.

I still took her to the sensory gym for some therapy. One year, the occupational therapist assigned to Talia there could not handle her behavior, and although the director was helpful to me in the past, she did not want to use an ABA approach to Talia's aggression and lack of motivation to engage in tasks. She complained that my daughter just wanted to stay in the occupational therapy swing and even complained that it was too much work for my mother to bring Talia there, although later she apologized for that statement.

I then decided rather than trying another center where I had no control over the therapist, I would do all Talia's occupational therapy at home. A good therapist or teacher will not see problems as simply the child's but rather as their own problems. They will find solutions so the child can develop his or her full potential. The home therapist worked well with the team and used an

ABA approach. This was also better because Talia would have more practice with ADLs.

I had installed a therapy gym in our basement. We already had a ball pit to provide tactile input, strengthen muscle tone, and provide body awareness. This can help with gross motor skills. In addition, after Talia tried it out, we also bought a steamroller, which is like a squeeze machine to provide deep pressure input and improve motor planning and processing. We purchased a textured balance beam and parallel bars for use by the physical therapist. The major problem was an occupational therapy swing. We bought a swing with some attachments, including a platform swing, a tire swing, and a cuddle swing after consulting with the director of the sensory gym where Talia had previously attended. At least, I hope it adds some value to the house.

## Aggression, Self-Injury, Pica, Food Stealing, and Improvement

In 2011, Talia started to become more aggressive and self-injurious. She started biting herself and others, including her sister Batsheva. I was worried I could not keep her at home. I told Batsheva not to be near her unless an adult was present. When we received an autism service dog, Cecil, Talia's self-injury and aggression decreased by about 80 percent.

However, I was disappointed when the synagogue we went to would no longer permit us to bring in Talia's service dog. After bringing the dog for a year, I was told by the rabbi's wife that another rabbi told him the dog should not be in the synagogue. The temple was convenient and around the corner. I explained that the dog stopped Talia from running into the street and helped with her self-injury and aggression so we could avoid residential placement, but to no avail. She suggested I alternate coming with my husband or come when Talia was with a therapist.

It was already so socially isolating to have an autistic child, and this just made it worse. The dog was accepted at a synagogue about a mile away from our home, but this was not convenient. In addition, the prevailing opinion in the interpretation of Jewish law is that service dogs are allowed in synagogues.

My husband and I thought about pursuing the issue under the Americans with Disabilities Act and state law, which make it a crime not to allow access to service dogs, but we would not want to go back to a synagogue with such animosity. I was so mixed up about it. The rabbi's wife had previously taken care of Batsheva and served her dinner when no one could meet her bus on a regular basis when Talia was at a center for therapy.

Even though the dog was helpful with Talia's behavior, Talia became much more dangerous again when she could not access the therapy swing. There were problems with the installation of the swing. The initial hooks came out. I then had it installed with a heavier eye hook, but that came out after a few months. Sometimes my husband or a therapist would forget when hooking up the attachment to use the rotational device, and when the swing rotated, it caused the eye-hook screw to rotate as well, which loosened it in the ceiling. We were lucky that when the swing fell down, Talia never got hurt. When the eye hook came out and the swing could not be used, Talia's aggression and self-injury increased tremendously, even though she had her service dog. The swing appeared to help her behavior.

Once, after a physical therapy session, Talia suddenly and quietly went up to my mother and bit her in the head. Talia was having thirty-minute-long rages. She would scratch, bite, head butt, kick, and squeeze continuously. She would hit and bite herself, and if interrupted, would turn her aggression on to the individual stopping her.

I was afraid to be alone in the house with her, and my husband and I started to contemplate the unthinkable-putting her in a residential placement. We decided to reinstall the swing. We had part of the ceiling taken down and an extra beam installed to support the swing. We also bought a tire swivel with a rotational device, which would not only stop the screws from rotating but also not strain the screws when she swung too far. The swing significantly decreased her aggression and self-injury, according to my data collection, and the swing has remained intact for four years. What a much better solution it is compared to medication management! Indeed, there are small studies and a literature analysis of eighteen studies that show that exercise can help with behaviors prevalent in individuals with autism and improve academics.[37] In addition, Talia learned to communicate with the use of an iPad, with a Proloquo2go program, so she can safely express her feelings and requests, pressing icons with pictures and written words and then imitating them, rather than exhibit a dangerous behavior.

Taking Talia on vacation is not really a vacation. She may not sleep well in the hotel room. Once before we had a therapy swing, we tried to take Talia away for an overnight to Longwood Gardens in Pennsylvania. Batsheva was really looking forward to this trip.

When my husband went outside to pack the car, Talia would run out and go to the backyard swing. My husband kept bringing her back into the house. Talia ended up crying nonstop the entire day, including the whole car ride, at the gardens, and into the evening. We were not sure if the swing was the problem or she was sick.

After driving more than an hour with a crying Talia, my husband suggested we go home. However, I remembered how disappointed I was as a child when my parents had to cancel trips because of Matthew's behavior, and I did not want to do that to Batsheva.

---

37 Lucy E. Rosenblatt, Sasikanth Gorantla, and John B. Levine, "Relaxation Response-Based Yoga Improves Functioning in Young Children with Autism: A Pilot Study," Journal of Alternative and Complementary Medicine 17, no. 11 (November 2011): 1029-35, doi:10.1089/acm.2010.0834; Fatimah Bahrami, Ahmadreza Movahedi, Sayed Mohammad Marandi, and Ahmad Abedi, "Kata Techniques Training Consistently Decreases Stereotypy in Children with Autism Spectrum Disorder," Research in Developmental Disabilities 33, no. 4 (July-August 2012): 1183-93, doi:10.1016/j.ridd.2012.01.018; Sabine C. Koch, Laura Mehl, Esther Sobanski, Maik Sieber, and Thomas Fuchs, "Fixing the Mirrors: A Feasibility Study of the Effects of Dance Movement Therapy on Young Adults with Autism Spectrum Disorder," *Autism* 19, no. 3 (April 2015): 338-50, doi:10.1177/1362361314522353; Leslie Neely Mandy Rispoli, Stephanie Gerow, and Jennifer Ninci, "Effects of Antecedent Exercise on Academic Engagement and Stereotypy during Instruction," *Behavior Modification* 39, no. 1 (January 2015): 98-116, doi:10.1177/0145445514552891; D. A. García-Villamisar and J. Dattilo, "Effects of a Leisure Programme on Quality of Life and Stress of Individuals with ASD," *Journal of Intellectual Disability Research* 54, no. 7 (July 2010): 611-19, doi:10.1111/j.1365-2788.2010.01289.x; Russell Lang, Lynn Kern Koegel, Kristen Ashbaugh, April Regester, Whitney Ence, and Whitney Smith, "Physical Exercise and Individuals with Autism Spectrum Disorders: A Systemic Review," *Research in Autism Spectrum Disorders* 4, no. 4 (October-December 2010): 565-76, doi:10.1016/j.rasd.2010.01.006.

We bought a thermometer, but Talia did not have a fever. The next day, she was fine. My husband and I both realized we should have let her play a few minutes on the swing before we left.

For a time, Talia had another serious behavior: pica, smearing and eating feces. My husband and I would often go into her room at night to find the carpet, bedding, walls, and even Talia covered with feces. In 2008, we found a worm and my whole family had to be treated for pinworm, and the house had to be sanitized. The chewable tablets did not taste good either.

Just as she was day trained to use the toilet, we had to night train her. My husband, a grandparent, or I would stay in her room while we waited for her to get to sleep. As soon as we saw her straining, we would firmly state "Bathroom!" and rush her to the bathroom. As soon as she went, we gave her a type of cookie that she could have at no other time. She decided she preferred the cookie to the fecal play and then started to go on her own, although once in a while she still had accidents. After a few months of having no accidents, we were able to stop the cookies.

We have also had embarrassing experiences in restaurants, although her school has helped and she has improved. She would not sit for too long. She used to dart away and take food from other customers. She once got away before I could catch her, and she stuck her finger in someone else's duck sauce and licked it. Once when my husband came back from a walk with her inside a restaurant, he discovered an eggroll in her hand, which she had taken off another table.

Her school, described later, has helped with her food stealing with a procedure called a differential reinforcement of other behaviors (DRO) and an interruption procedure. This involved keeping food in front of her at all times, and when she went a predetermined number of minutes without attempting to steal it, they gave her a small edible reinforcer. The school also put her hands down if she put something in her mouth. Another procedure the school has used is fixed-interval access, in which every predetermined number of minutes she received a very small edible, like a piece of popcorn. By feeding her something every few minutes, she was less likely to steal food. If she did steal food, the staff removed the food from her mouth and held her hands down for a few seconds. The school has also successfully and still uses the DRO for pica and mouthing objects. Talia now has a token board and earns tokens for not placing items in her mouth. She earns a token every few minutes, with the time intervals being gradually increased. After earning nine tokens, she can exchange them for a preferred reinforcer. If she puts something inedible in her mouth, there is a procedure called response cost where she loses her tokens and has to start earning them again. The school has also taught her to stay seated. She is now much easier to bring to a restaurant.

There have been other embarrassing experiences. My daughter loves zippers. One time on a fundraiser walk, she started to open the pocketbook of someone walking next to us. My husband and I apologized, but because it was an Autism Speaks walk, the other pedestrian told us we did not need to apologize. I no longer participate with any Autism Speaks events because their policy on aversives has caused problems for my family, as I discuss later. Now, Talia holds her service dog's handle with her right hand and holds a pocketbook with her left, keeping both hands occupied-skills she has been taught and is still being taught to do at school.

Furthermore, with the intensive behavior therapy, Talia learned self-control. Through dense positive reinforcement for doing her tasks and restraining her arms if she was aggressive, Talia learned to control her behavior when she was having a negative mood. Sometimes when she is upset, she will now start to grab someone, but before squeezing hard, she will stop herself.

## Special-Education School Placement

When Talia was three, my husband did a bike-ride fundraiser for the school, a Lovaas type 1:1 ABA school, as the McCarton Center and three of my colleagues at Mount Sinai felt Talia still needed. My husband was hoping she would be accepted. Unlike other private 1:1 ABA schools in New York, this one was state approved. However, the state only approved schools with a class ratio of 6:1:3.5 (six students, one teacher, three and a half paraprofessionals).

Because the state did not approve the school as a 1:1 instructional program, parents had a significant fund-raising requirement to maintain the program as a 1:1 ABA school. This is unlike the situation in New Jersey, where 1:1 ABA autism schools are funded as such.

The school would also have required us to live on Long Island as a condition for acceptance. This was because with a previous student from New York City, even after the impartial hearing, the school had to wait six months to receive tuition money. By New York City delaying funding for six months, they won something much better for them than the impartial hearing that they lost as the school would never again accept a New York City student.

We looked into where to move, consulted our attorney, and researched which school districts would approve funding for additional ABA therapy. We also did not want to move somewhere that Batsheva would have to switch schools. She was just adjusting to her new school, but I could only get busing within fifteen miles from a residence in Long Island. We looked at some houses in West Hempstead, but ultimately Talia was rejected by this school.

I then played what I call autism lotto. I applied to the New York Center for Autism, a New York City charter school, which had a Lovaas 1:1 autism program. The school informed me they had about seventy applicants a year, and seven would be accepted. The school also did not provide speech, occupational, or physical therapy.

Prior to the lottery drawing, I had dreams about it. When I attended the drawing, some the parents of the losing children were in tears. My child was one of the losers, but I did not cry.

I could have reapplied for the charter school the following year, but I took my mother's advice and did not. My mother learned the school only taught students up to the age of fourteen. My mother told me I would start worrying when Talia was twelve years old about where I would place her next, which was the same problem my mother had in placing Matthew after he was thrown out of Reece School. She said what I really needed was a school that taught students to the age of twenty-one.

Other 1:1 ABA schools in New York City were very expensive and had long waiting lists. My daughter was on the waiting list for almost two years for one of them. In 2007, at the age of five, Talia eventually got into the ELIJA (Empowering Long Island's Journey through Autism) School, a 1:1 ABA school.

I remember I was on the subway on the way home from work when I received a voice mail from my mother. New York City subways did not have cell-phone service between stations. Concerned there was a problem with my children, I called back when I reached Penn Station. My mother told me the ELIJA School called but had not left a message. I was so eager to find out if she was accepted that I could not wait until the next day.

When we learned she was accepted it felt so wonderful. I called my relatives immediately with the good news. It was better than getting into medical school; it was my daughter's life. It was a chance that maybe one day she would talk. However, unfortunately for most people in New York City, ABA-the most scientifically proven method by far to show significantly positive outcomes for treating autism[38]-is simply unavailable.

Although I wished the DOE could find an appropriate placement for Talia, I did write a letter to a state-approved private school, New York Child Learning Institute, asking them to consider Talia. If that did not occur, the next step for us was getting ELIJA funded. By federal law, to obtain funding for a program not recommended by the DOE, the district must fail to offer an appropriate IEP or program, and you must show that the program you selected was appropriate and that you have an open mind. To demonstrate you have an open mind, you must be open to what the department recommends, but that does not mean you have to agree to it. For example, it is important not to make a down payment to a private school unless the tuition contract states you will receive a full refund if you accept a DOE-recommended program.

---

38 According the New York State Department of Health, ABA therapy was the only treatment that showed effectiveness in well-controlled research and "recommended that principles of applied behavior analysis and behavior intervention strategies be included as an important element of any intervention program," for young children with autism, *Clinical Practice Guideline: The Guideline Technical Report, Autism/Pervasive Developmental Disorders, Assessment and Intervention for Young Children (Age 0-3 Years)* (New York State Department of Health, no. 4217, 1999), IV-25.

Otherwise, the school district can argue that the nonrefundable tuition deposit makes the parent unwilling to consider an appropriate district-recommended option.

## School Age Committee on Special Education Encounters

The School Age Committee on Special Education (CSE) attempted to test Talia at the age of four. I brought her down to a CSE building, where we were taken into a room with a conference table, adult-size chairs, and large windows. She looked so small in that adult-size chair, with her feet far above the ground. The only "toy" Talia could find was the phone cord, although she did appear to enjoy watching the cars out the window. The examiner attempted to test Talia, but the testing materials (Wechsler IQ) were way above Talia's cognitive skills, and Talia showed more interest in the window anyway. The examiner tried another room with a similar conference table and chairs, but Talia just looked out the window there too.

The examiner requested I leave the room during the testing, but my mother told me beforehand not to leave. She was concerned because Matthew once came out from a CSE evaluation appearing traumatized. The CSE examiner wrote in the report that he had used physical force to get him to comply with a task. My lawyer later told me that I should let the CSE know beforehand that my brother had a negative experience and that I was concerned and wanted to observe. A child can act differently from seeing a parent, which can interfere with an evaluation. My lawyer even told me not to show up at school on the date of a district observation.

Soon after this, I attended the IEP meeting a few weeks before Talia's fifth birthday. I received the meeting notice less than the five days required by New York State regulations, and I had kept the envelope to show the postmark as proof. This meeting was conducted without speaking to a single therapist, looking at therapist reports, or having any objectives or goals.

At the meeting, I discussed Talia being dangerous. I stated that she would climb up the back of a chair because she did not know that the chair might fall, that she mouthed objects, and that she could run away and go out into the street. Although I did not state this, we had to remove the dresser in her room and lock her closet because she was climbing up shelves. I stated that she would go to the bathroom if she knew where one was, but she would not tell you she needed to go when she was outside. I also stated that Talia played in the toilet bowl and needed help wiping herself and washing her hands. I was told that there was a bathroom in the classroom, and she would receive assistance when she used it. I was also told that Talia would be put on a bathroom schedule.

I was told that Talia would be classified as autistic, and I was offered a 6:1:1 class. The officials said that some children had their own paraprofessionals, and therefore typically the class was 6:1:2 and could be as high as 6:1:3 or 6:1:4. However, if those paraprofessionals would not be for Talia, then this ratio was irrelevant. I asked if 6:1:1 was the most intensive training available and was told that no programs were more intensive. Of course, they did not mention the autism charter school with 1:1 ABA because I did not win the lottery mentioned earlier.

Toward the end of the meeting, we discussed the possibility of getting Talia her own paraprofessional. I again brought up my concerns about Talia's safety and her problems maintaining attention. I was told that maintaining constant attention might not be possible and there needed to be downtime to play. I specifically asked if Talia could be kept safe without a paraprofessional. The psychologist said she could be. The officials said that a child's problems with danger were posted on the classroom wall. They said that two adults in the class were sufficient, and too many staff in the class would be distracting.

When we discussed speech therapy, the officials recommended to reduce it to what they said was the maximum, which was not true, and would take place in school. They said that since she would be getting language in school, it was justifiable to decrease Talia's speech therapy. They said any increase would have to be requested by the therapist at school.

At my IEP meeting, I was told that the methods used were ABA, TEACCH,[39] and the Miller method.[40] When I inquired about the Miller Method, I was told by the parent representative that it used platforms to teach safety and if I wanted more information, I should go online. I later took the parent representative's advice and searched online, only to find the Australian

39 *Treatment and Education of Autistic and Related Communication Handicapped Children* (TEACCH), "TEACCH developed the intervention approach called 'Structured TEACCHing,' which is based on understanding the learning characteristics of individuals with autism and the use of visual supports to promote meaning and independence. TEACCH services are supported by empirical research, enriched by extensive clinical expertise, and notable for its flexible and individualized support of individuals with Autism Spectrum Disorder (ASD) and their families... PRINCIPLES OF STRUCTURED TEACCHING:

- Understanding the culture of autism
- Developing an individualized person- and family-centered plan for each client or student, rather than using a standard curriculum
- Structuring the physical environment
- Using visual supports to make the sequence of daily activities predictable and understandable
- Using visual supports to make individual tasks understandable,"
  *University North Carolina School of Medicine TEACCH Autism Program*, accessed January 10, 2016,

https://www.teacch.com/about-us/what-is-teacch.

40 "A description of the program can be found on the Miller Method® website (Miller & Eller-Miller, 2006). The Miller Method is based on 'Cognitive-developmental systems theory' which assumes that typical development depends on the ability of the children to form systems and organised 'chunks' of behaviour. The program claims to transform the child's 'aberrant systems (lining up blocks, driven reactions to stimuli, etc.) into functional behaviours.' Strategies employed include narrating the children's actions while they are a metre above the ground on an 'Elevated Square.'" Jacqueline M. A. Roberts and Margot Prior, *A Review of the Research to Identify the Most Effective Models of Practice in Early Intervention of Children with Spectrum Disorders*, Australian Government Department of Health and Ageing, 2006, 73; Arnold Miller and Eileen Eller-Miller, "What Is the Miller Method?" retrieved May 25, 2006.

Government Department of Health and Ageing called the method "preexperimental" in nature due to the lack of controlled research.[41]

All these services were cut and changes made despite not speaking with the therapists, reviewing any reports, or discussing any goals. At the end I was told I could not get an IEP because they had no goals on file. They asked me to have my providers fax them.

Three weeks later, without an IEP, I received a recommended placement. I again kept the envelope in case I needed proof that I received a placement recommendation without an IEP. The postage cost on the envelope would have been much higher if a usual multipage IEP was inside.

Years later, when Talia was nine years old, during an IEP social history, a CSE social worker asked me if I played music when Talia was aggressive to calm her down. I explained to her that music could serve to reinforce the behavior because Talia would engage in it again to get music. It would also remove the motivation to work for music, and if Talia was allowed to stop what she was doing to get music, she might engage in the aggression again to escape a task.

During a meeting later that day, when we discussed behavior, the ELIJA School discussed that they took data in partial two-minute-interval collections, which meant that every hour was divided into two-minute intervals, and the observers would check off behavior if it occurred during that time. The CSE psychologist and the social worker did not want data in that format,

---

41 "Jordan, Jones and Murray (1998) conducted a review of research evidence for the effectiveness of the Miller Method. They reported only one study of outcomes of the program, which failed to evaluate the direct effects of the independent variable (i.e., the treatment program). Further research is required in order to evaluate the effectiveness and long-term outcomes of the program. Jordan, Jones, and Murray (1998) cautioned that, in the absence of such research evidence, the program must be considered pre-experimental in nature." Jacqueline M. A. Roberts and Margot Prior, *A Review of the Research to Identify the Most Effective Models of Practice in Early Intervention of Children with Spectrum Disorders*, Australian Government Department of Health and Ageing, 2006, 73; Rita Jordan, Glenys Jones, and Dinah Murray, *Educational Interventions for Children with Autism: A Literature Review of Recent and Current Research*, Department for Education and Employment, no. 77 (Norwich, UK: Crown Copyright Unit, 1998). According to the Miller Method website, http://www.millermethod.org/, *The Language and Cognitive Development Center (LCDC)*, accessed January 15, 2016, there have been no published research clinical trials on the Miller Method since 1998.

despite the fact that partial interval recording is one of the accepted procedures for recording behavior data in applied behavior analysis.

After the meeting, I received the IEP in the mail. Without any mention of it at the meeting, the CSE had removed most of the ADL goals, such as showering and changing a sanitary napkin. They did this despite my answer to the psychologist's question about what was most important to me: I replied it was the ADLs. I stated that when my daughter would be living in a group home one day, I did not want some stranger scrubbing her down, which the psychologist said was understandable. The IEP also stated she would have ten seconds to respond in some trials, although that would be too long and was never discussed with me at the meeting.

At a different meeting, a year later, a psychologist told me she believed home ABA was overkill because Talia's needs were addressed at school, so there was no need for home ABA. I objected. She asserted that Talia's learning to feed herself could be addressed by the occupational and speech therapists. The home ABA supervisor mentioned that Talia used more independent skills at home, and if one variable was changed from a program, Talia did not generalize. Then the psychologist stated that if what Talia learned at school was not generalized, then the school was not effective. She reiterated that it was repetitive for Talia to have therapy in both places. My mother, who was at this meeting, replied that the home program reinforced what Talia was learning in school.

I mentioned another reason Talia needs the home program: she regressed easily. For example, over the three-week summer break, she regressed tremendously with her showering program, which was not being done at home.

The psychologist stated Talia needed a school with a longer school year. I asked her which school had longer than a six-week summer session. Schools are legally required to have six-week summer sessions over a ten-to-eleven-week summer school break for children who would regress over a two-month break, and the state does not fund more, so no day schools in New York have a longer summer session. She never answered my question.

I asked about putting Talia's service dog (as discussed earlier) on the IEP, because the dog prevents elopement and helps for social goals. I explained that

Talia could wander, but she knew to hold onto the dog handle. I told her Talia had a belt with a connection to the dog to further prevent elopement. The CSE psychologist suggested that Talia hold an adult's hand or use a rope and that she does not need a dog. The school's associate executive director mentioned that holding an adult's hand did not promote independent walking.

Another year later, at a different IEP meeting, the psychologist told me they would drop showering and full dressing because their recommended schools could not do it. I stated I did not agree with this decision. The special-education teacher stated that parents should be working with their children at home, and I should be teaching her myself to shower at home with some parent training. I told them I did help with maintenance, but that skill acquisition was different. I noted that typical children go to school, and we do not teach parents to teach academic skills at home. I told them that we did not expect a parent to teach their child geometry at home, that I was not a special-education teacher or a board-certified behavior analyst (BCBA), and I did not have the credentials to teach Talia these skills at home. The psychologist mentioned learning dressing on a doll, and I told her Talia would have a generalization problem.

## Why Some School Placements Were Inappropriate and How to Fix Problems and Get Results

At some placements, students engaged in self-stimulatory behaviors such as rocking, jumping, hand flapping, mouthing, and making loud noises. At some placements, there were no DROs to address self-stimulatory behavior that interfered with learning and could even be dangerous, such as spinning and head banging-which Talia has-and reinforcement inventories were sometimes done infrequently, even only twice a year. Also at some placements, staff ABA training was too low for staff to employ ABA procedures according to research evidence-based methodology, and in one place in particular, staffers were only trained twice in ABA, and there was also no regular ABA supervision. At some schools there was no individualized parent training, despite research evidence presented elsewhere in this book. In addition, at some autism schools, there was no explicit training on social skills goals, despite that socialization is a core impairment in autism.

Schools did use TEACCH icons and Picture Exchange Communication System (PECS) books.[42] At one placement, I told the staff that the way I was trained in the TEACCH program, which was used at one of my jobs, was to assess what the individual understood. Sometimes a person needed an actual photo or object.

I would also like to note that programs I visited were not comprehensive TEACCH programs. The TEACCH program was designed in North Carolina, and it has continuity of care in the home environment with parent training and even in the construction of materials to use at home. Also, I have noticed that the PECS icons were not differentiated from the TEACCH icons (i.e., different borders), both using the same Mayer Johnson symbols, so children could discriminate between them.

Another problem was that, like ABA programs, many TEACCH programs, including those I have visited, do not do what they are supposed to: they are not used the way they were designed and researched, the way research showed that children make progress. For example, the programs require teaching prerequisite skills such as matching and discrimination, including distinguishing a picture from a background. It is useless to put symbols of the activities on the schedule if the student does not distinguish the symbol from the background, does not understand that a picture represents an object, or if the student is not able to match a photo to an object.[43]

---

42 "Developed by Andrew S. Bondy, Ph.D. & Lori Frost, M.S., CCC/SLP…PECS was developed in 1985 as a unique augmentative/alternative communication intervention package for individuals with autism spectrum disorder and related developmental disabilities…PECS begins by teaching an individual to give a picture of a desired item to a 'communicative partner,' who immediately honors the exchange as a request. The system goes on to teach discrimination of pictures and how to put them together in sentences. In the more advanced phases, individuals are taught to answer questions and to comment." *Pyramid Educational Consultants*, accessed January 10, 2016, http://www.pecsusa.com/pecs.php.

43 Fry Williams and Lee Williams. *Effective Programs for Treating Autism Spectrum Disorder: Applied Behavior Analysis Models*, 202-3. Lee Marcus and Eric Schopler, "Educational Approaches for Autism, TEACCH," in *Clinical Manual for the Treatment of Autism*, ed. Eric Hollander and Evdokia Anagnostou (Washington, DC: American Psychiatric Publishing, 2007), 220-21.

Furthermore, even if the student does have those skills, the student may not be able to recognize and discriminate those symbols if there are too many of them in the field. Yet schools will line up the same symbols for each student, one next to the other, not knowing if the student really understands any of it.

This shows a lack of individualization, which is part of a true TEACCH program. A good TEACCH program uses individualization and teaches prerequisite skills. If a child does not understand a two-dimensional representation of an object, the instructors develop an object-activity schedule.

In a true TEACCH program, the schedules are accessible to the student's view. The schedules are not placed in a part of the classroom that students do not usually see. The backgrounds are designed to interest the student. If a student likes trains, for example, the schedule can be designed on a train background. If a student likes a particular color, the background could be that color. If the student cannot recognize multiple symbols in a field, the schedule could be made in a photo-album format, with one icon per page.

I have witnessed in practice how a well-done visual activity schedule can help with agitation and aggression. For example, if anxiety, agitation, and resultant aggression occur around transition times, the staff can recognize the problem. They can then work to make the transitions predictable using the schedule, which can sometimes avoid the need for psychiatric medications for behavior.

For Talia, we found a visual schedule in her closet with pictures of clothing items on different shelves (for example, a picture of a shirt on a shelf with shirts) and identical pictures on a schedule on her closet door have helped her dress independently. These steps are known as task analysis. First, my daughter will check her schedule, take the picture of underwear (for example), and match it to an identical picture of underwear on a shelf. She removes the underwear from the shelf and puts it on. She then does the same for other articles of clothing.

Similarly, a photograph-album schedule has taught her to set the table. By turning to the page with the picture of a plate, she learned to put a plate on the table and then do the same for a fork and cup. Now, she is able to set the table without the visual schedule.

In addition to those skills, Talia has learned hand washing and showering skills with discrete steps. She has also learned to use bobby pins in the front of her clothes to differentiate front and back. Her dentists have told me what great hygiene she has, and one even said she wished her child patients without problems would brush that well. That is because through ABA she learned to tolerate brushing, and with the task broken down into twenty-three small steps, she has been brushing more independently.

In addition to schools not applying scientific evidence-based approaches to treatment, I observed other problems. At one placement, sometimes speech therapy was in groups of two-even if the IEP stated 1:1. Despite being an autism program and the high comorbidity of autism with verbal apraxia, which Talia also has, no speech and language therapist was trained in apraxia.

At one program I visited, occupational and physical therapies were done in the hallway, and speech therapy was done mostly as a push-in, but also in the hallway at times. Some children with autism such as Talia can get distracted easily and may not make progress when receiving therapy in these environments.

At one placement, students were taken to the bathroom every two hours. The problem with this was if students such as Talia were on a fixed schedule, they would not learn to request to use the bathroom.

I have witnessed children at placements eating off the floor or engaging in pica, such as eating plastic and not redirected. Sometimes, children were in noisy lunchrooms despite that individuals with autism, including Talia, can exhibit agitation, aggression, and self-injury when exposed to loud noises.

At one school, one student picked something off the sidewalk, and no staffer removed it from his hand. Sometimes, staffers handled food without gloves on. At one school, the coordinator told me that she would not get the IEPs when school started.

At another program, the special-education teacher was absent. The substitute in the room was only a paraprofessional, which is a violation of federal law. At another placement, paraprofessionals were absent without substitutes, another violation of federal law.

At a different school, the education director said things that made no sense and were contradictory to published research, including that completing sentences in songs was a nonfunctional goal (although it is a goal on the revised Assessment of Basic Language and Learning Skills by James W. Partington, who used a research-based verbal behavior approach) or that everyone got used to a thirty-minute interval to earn a token reinforcer. That also is contradictory to ABA research. Someone may be willing to work a minute for a cookie, but not for half an hour.[44]

### *Individualized versus State Curriculum*

At one school, the school psychologist told me that they followed the state curriculum. Students were working on learning what people ate in medieval times, how people lived then, how people in castles put tapestries on walls for warmth, and students learned about dragons and fairy tales. During science lessons, the class was working on climate and rocks.

This would have been a total waste of time for Talia, who could not even identify real animals at the zoo. According to one article, "[t]he programs that are effective have a planned, intensive, individualized, and flexible curriculum according to the individual needs of the child."[45]

One school, to my surprise, immediately rejected Talia without my ever visiting. All I did was share about some of Talia's behaviors over the phone. I told them she bites; at her current school, she was kept five feet away from the other students; she swallowed a dime in the last year, had to be x-rayed, and then swallowed a toy that came out in her stool two days later. I said that I did not know how she swallowed the toy, but she must have waited to do it when no one was looking. Finally, I said she was a runner and had a service dog with her at all times, including school, to prevent that. The social worker told me their placement would not be secure enough, and the school would reject her.

---

44 In fact, "[a] *progressive-ratio schedule* reinforcer assessment helps to test if a preferred stimulus will continue to be an effective reinforcer as the response requirements become greater." Fry Williams and Lee Williams, *Effective Programs for Treating Autism Spectrum Disorder: Applied Behavior Analysis Models*, 191.

45 Muhammad Waqar Azeem, Nazish Imran, and Imran S. Khawaja, "Autism Spectrum Disorder: An Update," *Psychiatric Annals* 46, no. 1 (January 2016): 60, doi:10.3928/00485713-20151202-01.

One program accepted Talia, and then the Central Based Support Team (CBST) sent a letter to the Committee on Special Education (CSE) to have an IEP meeting for that school, even though no one at CBST ever evaluated her or was at the IEP meeting two months earlier to determine her needs. Yet, they usurped the IEP committee's authority on determining placement. For some children, the CBST will even attempt to override a CSE IEP committee that recommends a residential placement. The parents then have no choice but to file for an impartial hearing.

After visiting one placement in particular, which I felt was completely inappropriate for Talia, I made the decision that if I ever lost a hearing for that school, I would rather work more than one hundred hours a week, like I did sometimes during my residency training, to pay the tuition for Talia's school than send her to an ineffective program. I still worry about losing hearings and having to work long hours to pay for her care. I worry about Batsheva, too, because if I have to work eighty or more hours a week, I would not be around for her.

## Personal Debt and Anxiety Fighting for Effective Autism Treatment

Ultimately, in 2007, we did not agree with Talia's initial DOE-recommended placement and had our attorney submit an impartial hearing request. Almost two years later, we were significantly in debt after paying for two years of school tuition, home therapy, private evaluations, and legal expenses for our child-without any reimbursement. The DOE had paid for only some of her occupational therapy and six hours of speech therapy. Otherwise, we were on our own.

The government always tries to find more ways not to pay. The first year my daughter attended the ELIJA School, my mother-in-law wrote the checks for tuition. When my husband and I signed our stipulation (a written agreement with the DOE to reimburse most educational expenses in return for dropping the impartial hearing request), the DOE sent my lawyer a reimbursement check. Now the city will not reimburse the parents if grandparents or anyone else writes tuition checks without the parents producing a notarized loan agreement. It took three years to get reimbursed for speech, occupational, and physical therapy.

My husband and I have borrowed money to pay for Talia's school and her therapists. Between my residency training and my autism fellowship, I only worked about seven hours a week, because I wanted to get home with my children. Therefore, we had little savings, and Talia's needs precluded me from full-time employment. We never thought we would have to watch our money so carefully. We constantly have had to worry about money and have gotten into arguments over minor purchases, which we would not have done in the past. I have had nightmares about not having money to pay my expenses.

When I finished my autism fellowship, I was temporarily out of work. I had a job lined up, but the human resources department mistakenly told me the position came with malpractice insurance. As a physician, it would be dangerous to practice without insurance, so I found another job, but I had to wait for my malpractice insurance to be approved first.

People told me to look at my time out of work as a break, but I could not enjoy it at all. It was no break when I went to buy food and I had to look for the cheapest available. I had to watch every nickel and dime.

I was buying chopped meat because it was cheap. My mother asked me why I was buying it, because it is unhealthy. I told her I was uncomfortable spending the money, and then she started buying me meat.

While I was out of work and struggling financially, I did have one pleasant distraction: I cooked lunch for the staff at Talia's school. The parents took turns bringing lunch to the staff, but instead of buying restaurant food, I prepared inexpensive vegetarian dishes, baked cupcakes, and brought it down. It took me three days to prepare everything, but the staff really seemed to appreciate it.

We went four years without taking a vacation, and then my husband and I only went away to Harrisburg, Pennsylvania (no offense to Harrisburg, but it is not known as a tourist hot spot). We went for four days, including one day when we went to Shenandoah National Park. Spending a day in a cave and hiking a trail was great and relaxing. I usually have no problem falling asleep, except close to IEP meetings and impartial hearing dates, but often I will wake up tense and anxious between three thirty and five o'clock in the morning, unable to get back to sleep. On vacation, I never have this problem.

We were buying second-hand clothes and never went out to eat. I remember going to work in an old and dingy winter coat. When my husband bought me a new one, rather than being thankful, I was upset about the money he spent.

My husband was going to court in the summer in winter suits because we had no money to buy summer suits. He collected cans and bottles at work to return to bring in a little extra cash. Once when a stipulation of settlement was signed, my husband brought home some sushi to celebrate, but I thought maybe we should have waited until the check arrived. I joked around that the reimbursement check was *always* in the mail, but my husband did not think that joke was funny at all. I can understand his view. I developed a new appreciation for what my mostly poor patients' families go through.

When I finally received an authorization pay memo for home ABA, I submitted my invoices and canceled checks to the Bureau on Nonpublic Schools Payables (BNPSP) in downtown Brooklyn. We did not receive immediate payment. After I finally spoke to someone there, I was told they never received my paperwork, even though I had sent it certified with a return receipt.

I went down to their office with copies and the return receipt. When I drove down, I was uncomfortable paying for a garage, so I spent over half an hour to find street parking about a mile away. I ran down to the office because I had a doctor's appointment a little later. I was out of breath, and the person inside told me to come in and gave me a chair. She told one of her colleagues that she might need to call 911. After resting a few minutes, I showed her the return receipt with the correct address. I was told no one by that name worked there. I dropped off the copies, and the person in charge of reimbursement later called to tell me they did not accept copies of invoices. I argued that I had mailed the originals but that the office claimed they had never received them. She ultimately reimbursed us off the copies.

After that our ABA providers were supposed to receive direct payment for services rendered, although for one provider it took more than a month to validate her Social Security number. When I investigated, the person who worked in an upstairs office at BNPSP told me she submitted the request to someone in the same building but in a downstairs office. When I called the person in the downstairs office, she told me she never received it. I then called again the person in the upstairs office, who insisted she sent it but that the person in the downstairs office did not know what she was doing.

They later did not pay me on one of the invoices, and I had to call to point out the mistake, and then they overpaid me by fifty cents. Years later, the BNPSP held up reimbursement for months claiming they wanted exact dates on the notarized affidavits, even though the exact dates were already on the notarized invoices, and they had been submitted in the same way previously for years. They did not contact my lawyer or me about the details they now wanted but rather just held off on reimbursement until my lawyer's office called after two months to find out why the check never arrived. Again, the city was always looking for ways to avoid making payments.

### *Getting to and Home from School*

In 2007, when Talia first started at the ELIJA School, busing was not set up, and I had to drive her to Long Island and then go to Manhattan to finish my autism fellowship. Because I did not get to work until ten thirty or eleven o'clock, I would have to work until seven o'clock in the evening, and I would not get home until eight thirty.

One year, the CSE never performed an IEP. However, in the summer I could not get her busing without an IEP. I could not drive her to school every day because I had to be at work by nine o'clock. My mother did some of the driving, but she was also out of town for a few weeks. We paid $110 a day for transportation to take Talia to and from school. My father did not drive, so he sat in the back seat with Talia-we could not leave Talia alone with an unknown driver. My father stayed at my house and stayed at the school for the day until it was time for Talia to come home.

### *Being My Own Plumber*

Once when we were financially desperate, the water filter underneath our kitchen sink was leaking. Being uncomfortable with spending money on a plumber, I decided to try to change the water filter myself. It looked easy enough. To make a long story short, I accidently broke a pipe and ended up flooding our entire kitchen. I had to call an emergency plumber. My husband later said to me, "You're Jewish, and you're a professional. You don't do plumbing."

### *A New Oven, Just in Time*

Our oven was in such disrepair that we used fireplace matches to light the stove for years. We had finally been authorized for some reimbursement from the DOE when my husband noticed a sale at Home Depot. I knew the processing time after the signing of the stipulation would be four to six weeks. Although he was concerned we did not have the check in hand yet, I explained that even if we were late on one credit card payment because we were not reimbursed in time, we would still save money because the oven was on sale. I found out if we applied for a Home Depot credit card, we would not be charged interest for two years. We could delay a payment without cost. It turns out we did receive the check couple of weeks later.

The day before the oven was to be delivered, I had the plumber unhook the gas line. That evening, when my mother opened the old oven to retrieve the broiling pan, part of the door came off on her. We were lucky the oven was not on, or she would have been burned. The reimbursement authorization truly came just in time.

### *Mayor Michael Bloomberg and Nightmares*

When Mayor Bloomberg was running for his third reelection, he sent campaign people door to door. When the messenger arrived at my house and asked me what I thought of Mayor Bloomberg, I told him to wait and that I wanted to show him something. I took out over two years of unreimbursed invoices and canceled checks-enough papers to fill up a large book-and showed it to him. I asked him why other students from Long Island at my daughter's school were getting their tuition and other services directly paid for by their school districts, but I had to go through this. He simply stated, "I can't imagine what you must be going through." He left without even trying to get my vote. I cannot help but think that when we owe the government money, they are not patient. The only time I have uncontrollable tremors is when our lawyer is calling me to discuss possible hearings or settlements with the DOE.

In 2006, Michael Bloomberg was the mayor, and he hired consultants, without competitive bidding, with a $17 million contract (later changed to $15.8 million) to help the DOE save money.[46] Even though the DOE paid these consultants millions of dollars, the city became very litigious to fight parents such as myself in order to avoid paying for needed services. Once New York City offers a settlement amount, it has to be approved by the comptroller's office. Our then Republican mayor and Democratic comptroller crossed party lines and collaborated quite well together against children with special needs.

Though the city agreed to pay for most of the costs, I was still responsible for some of the funding in addition to legal expenses and payments for private evaluations. After our agreement was settled with the New York City DOE to reimburse most of the money, the agreement sat in the comptroller's office. The stipulation took almost one year to sign. The previous year's agreement took six months for the comptroller's signature. In 2009, when I finally had my stipulation signed, I had almost two years of back invoices and canceled checks for speech, physical, and occupational therapies.

I spent four days at the Committee on Special Education (CSE), and together with a special-education teacher, we organized invoices and cut and taped canceled checks by discipline. He even once worked from the morning until one forty in the afternoon, when he had to start another meeting, without taking a lunch break. He even gave me his cell phone number and told me if I had a problem, I could still call him, even if he was on break.

---

46 Erin Einhorn, "Bloomberg Defends Fat Ed Rehab Tab," *New York Daily News*, August 30, 2006, http://www.nydailynews.com/archives/news/bloomberg-defends-fat-ed-rehab-tab-article-1.609544; Elizabeth Green, "Concerns Grow Over School Consultants' Performance," *New York The Sun*, October 16, 2007, http://www.nysun.com/new-york/concerns-grow-over-school-consultants-performance/64616/.

I was surprised by his kindness. However, I was not surprised when I was informed that the DOE mistreated him because he was helping children obtain certain essential services. The city's seemingly blanket policies for children attending nonpublic schools is a violation of federal and state law.

When he was gone, I told my mother, "There is no one I can trust at the CSE." My mother replied, "There was no one I could trust at the CSE ever!" My mother always dreaded meeting with the CSE, and growing up, I remember her repeating that the CSE would throw your child into the river. I remember her often worrying about my brothers not getting their placements funded and that she could not possibly afford their tuition. It was not helpful that my father was pessimistic and often told her she could not win. Many times that I met with the CSE, I was tense and generally could not sleep well before IEP meetings. It has now become an IEP-meeting annual tradition to get out of bed about five o'clock on the morning of IEP meetings because I cannot sleep anyway, go outside and run five to six miles to calm down, and then get my children up and ready for school. The IEP process literally gave me nightmares.

Here is an example of one of my nightmares:

> It was soon after the BP oil spill. I had moved to Louisiana, and it was at night. There was a grassy field full of oil and mud, and people were walking in all different directions. It was complete chaos. I was in the middle of all this chaos at a picnic table with countless papers stacked on the table in a completely disorganized fashion, about to have an IEP meeting.
>
> I handed my old IEP to a CSE worker. I did not want to lose Talia's services from the old IEP. I was told the meeting would be at another table across the field. I walked over to the other table, and we were about to start, but when I inquired no one had the old IEP. I told them I gave it to someone at the other table. I told them to wait, but someone told me they would not wait and that it was being started. I ran though the muddy oily field to the previous table, looking through the stash of countless documents, searching in the dark

for the old IEP, having trouble seeing the writing on the papers in the dark.

In September 2011, I called the CSE to obtain paperwork to get Related Service Authorizations (RSAs) so Talia's speech, occupational, and physical therapists and counseling providers could get directly paid by the DOE. I was told that now I could not get the paperwork without a court order because Talia was a Carter case; that is, a non-state-approved private school with the parent seeking tuition reimbursement. The DOE was trying to not provide Talia services she vitally needed.

A week later, I showed up at the CSE with a court order. After requesting the RSA, the chairperson, whom I never met beforehand, decided to meet with me and asked me questions about Talia. She informed me she wanted to develop a new IEP for her. It was extremely stressful.

After my encounter with the chairperson, I bought fish at a store and told the clerk I would bring my car over to pick it up. I had trouble finding where I parked my car, though, and then I forgot to pick up the fish. It was only when I entered a supermarket close to my house that I remembered the fish, and I had to go back to pick it up. I later realized I accidentally bought a different variety of fish than I had intended, a variety I knew my husband would never eat.

The day after this encounter, I was awake at four o'clock in the morning, all tense after waking up from a nightmare. I thought about a medical education article I once read about rescripting nightmares, about how to change the ending into something favorable. I then fantasized telling people at the CSE how I taught doctors in China how to diagnose and manage autism and speaking with terminology above their heads and totally intimidating them. After a few minutes, I was able to fall back asleep, at least that time. Other times the rescripting techniques did not work.

Besides nightmares, some parents will have symptoms of posttraumatic stress disorder (PTSD) over the battles obtaining needed services for their children. I know I get anxious when the mail arrives, worrying it will be some notice from the CSE. Symptoms of increased arousal in PTSD-such as irritability or outbursts of anger, poor sleep, exaggerated startle response, hypervigilance, problems with concentration, and mood symptoms, such as depression-can easily occur when parents are going through an impartial hearing process or an encounter with the CSE.

Having a special-needs child in itself can induce anxiety and depression. The battles to obtain services add another layer that may be even more stressful. Although symptoms of PTSD occur, the disorder cannot be diagnosed as "post" because the parent is currently going through the situation.

After one IEP meeting, I went home and was very tense. I started typing my mother's and my handwritten notes. My lawyer had told me before the meeting to take detailed notes. While I was typing, the lawyer for Matthew called to prepare me for the meeting I had (along with other JRC families) four days later with the OPWDD. I picked up a notepad to take notes and inadvertently stuck my mother's IEP meeting notes into the notepad. I went out of my mind trying to find my mother's notes that evening. I even searched the garbage. I kept repeating in my mind the lawyer's instructions to keep good records and kept thinking this mistake could cause me to lose a hearing and could cost me a ton of money. When Shabbos arrived, while people were eating, I could not help looking for the notes. My husband and one of our guests also helped me search for the notes. My husband poured me some Amaretto to calm me down. The next morning, although I did not have the best sleep, my IEP meeting anxiety somewhat dissipated, and I could think more clearly. I thought about the notepad and, with much relief, found the notes.

I once had a nightmare that I was with Stuart and Talia on a crowded subway platform at night:

> They were both agitated, and there was an announcement that there would be a forty-minute delay. Stuart was about to lunge off the platform onto the track, but with my right arm, I held him close to me to keep him safe. But then Talia moved forward, and I tried grabbing the back of her clothes with my left hand, but my grip was not good, and she kept moving forward and was about to fall onto the track. I yelled for help, but no one was paying attention.

After meetings, I know I will be hearing from some placements. Every time my phone would go off with an unidentified number, I became so anxious. At work, I would try to let it go until the end of the day, but I wonder who called.

Twice I got nervous over nothing: one call was simply trying to sell me something I had ordered previously, and one was a videographer doing a tape for my daughter's school. When I checked those messages while leaving work, I realized how I got anxious and worried unnecessarily.

It is also interesting how so many students have not had school placement, including some in my practice. The DOE knows I will spend the money to pay for a private non-state-approved school and sue for reimbursement if they do not find a placement, so after almost every IEP meeting, a school contacts me.

## Impartial Hearings

One night in 2012, my lawyer informed me there was a strong likelihood I would be going to a hearing. I started to think about how I could manage working eighty hours a week to pay for Talia's treatment if we were to lose. That night, I had a nightmare that Talia went down a slide into a flooded street with waist-high water, which I was worried was infected, and I could not get to her. Starting two weeks before the hearing, I had nightmares that would wake me up every morning between three thirty and four thirty. In one of them, I had to work until three in the morning to make money for Talia's therapy, and then I went to Penn Station to catch a train home. There was no train, and I was all alone and worried that I would get mugged. I was wondering why I was even going home since I would have to be back in the office in a few hours.

I remember thinking that even though going through a hearing process is scary, I could not let an evil political environment triumph over my daughter's life; I would feel I was a failure as a parent. I then started to repeat to myself that I would win this hearing. I pretended that my anxiety was baggage, and I was throwing it away. The nightmares decreased, but I was still getting up in the middle of the night, my emotions shaken and my body tense. It took hours to fall back asleep. Needless to say, my husband and I were not getting busy.

Just before our scheduled hearing date was Rosh Hashanah, the Jewish New Year. This holiday has a liturgy of significant omens, words we say to wish for favorable events before we eat ritual foods. In the nearly twenty years I have been married, my husband would never touch beets until I read the omen requesting that "our adversaries be removed." My husband took the smallest beet slice I had prepared, swallowed it as quickly as possible, and drank plenty of liquid. The hearing was adjourned, and the impartial hearing officer recused herself.

Do not think for a minute that impartial hearings are necessarily impartial. First, by federal law, if there is a dispute over a service or placement in an IEP, the prior IEP must apply, which is known as pendency. To impede abiding by federal law, New York City refuses to continue the previous placement or preagreed upon services unless the parent has a court order from a pendency hearing. Special-education lawyers charge for their time, and even if pendency is granted, the legal costs often are not reimbursable, unless the child wins services at the final impartial hearing or the costs are included in the settlement.

At one time, Talia's home ABA services had been on her IEP for two years, but were not written on the subsequent IEP, so my lawyer requested a pendency hearing. The "impartial" hearing officer refused to schedule a pendency hearing, forcing us either to pay for Talia's services out of our own pocket or forgo them. Then this impartial hearing officer blamed the administrative office for not scheduling the pendency hearing. My lawyer disclosed that she had e-mails to prove otherwise, after which the supposedly impartial hearing officer recused herself. Our lawyer accidentally missed a service for ABA home supervision on my daughter's impartial hearing request and so had requested an amendment, which was approved by the CSE, only to be rejected by that same hearing officer, and only a few hours before recusing herself.

Impartial hearing officers in New York are trained by the New York State Education Department. For years, one of their attorneys was living with the state review officer. Appeals of impartial hearing decisions would go to the

drugs (meaning used for a different indication or dosing than a medication's FDA approval) to manage those behaviors, which is particularly risky.[51]

While our politicians put on shows, children have dangerous side effects to their medications and are not learning. It is a politically inconvenient truth that schools are often not properly serving children with special needs.

Education has experienced budget cuts. Imagine if there were such budget cuts to, for example, antibiotics. Imagine if an insurance company or the government was trying to save some money and only paid for half a dose of antibiotics if someone had an infection. Not only would it not be helpful but the patient might also end up with a resistant infection. The expenditure for half a dose would be a true waste of money.

In education, our society often does the same thing. There is not enough funding. The children are in classrooms with too many other students. Students with special needs not only do not receive an education but also regress or even reverse in their behavior modification if the program is not supervised properly. For example, if the teacher or paraprofessional says to a screaming child, "Be quiet, and I will give you some candy," the child now learns to be loud first, and the offer of candy will soon follow. To spend money halfway on education in which no learning is going on, and possibly even harm is being done, is also a true waste of money.

---

51 Julie M. Zito, Albert T. Derivan, Christopher J. Kratochvil, Daniel J. Safer, Joerg M. Fegert, and Laurence L. Greenhill, " Off-label Psychopharmacologic Prescribing for Children: History Supports Close Clinical Monitoring," *Child and Adolescent Psychiatry and Mental Health* 2, no. 24, published online September 15, 2008, doi:10.1186/1753-2000-2-24; "Nearly 90,000 adults go to emergency rooms each year for side effects of psychiatric medications," Rob Goodier, "Ten Drugs Cause Majority of Adult Hospitalizations for Adverse Psych Med Effects," *Psych Congress Network*, July 15, 2014, http://www.psychcongress.com/article/ten-drugs-cause-majority-adult-hospitalizations-adverse-psych-med-effects-18180; Researchers suggested that side effects might be reduced by cutting back on off-label prescribing and that "[r]oughly two-thirds of prescriptions for typical and atypical antipsychotics are off-label and it is likely that many of those led to many of the emergencies, the researchers write." in "Ten Drugs Cause Majority of ER Visits in Adults for Adverse Psych Med Effects," *Dominion Diagnostics*, accessed October 9, 2016, https://www.dominiondiagnostics.com/news/ten-drugs-cause-majority-er-visits-adults-adverse-psych-med-effects; Lee M. Hampton, Matthew Daubresse, Hsien-Yen Chang, G. Caleb Alexander, and Daniel S. Budnitz, "Emergency Department Visits by Adults for Psychiatric Medication Adverse Events," *JAMA Psychiatry* 71, no. 9 (September 2014): 1006-14, doi:10.1001/jamapsychiatry.2014.436.

As I noted earlier, Talia has multiple behavior problems, primarily aggression that includes kicking, head butting, squeezing, pinching, scratching, hitting, and biting herself and others, as well as head banging and pica. I take time to document these behaviors, along with what occurred just prior, with whom, at what time, and what I did afterward. I show these records to the school and a board-certified behavior analyst (BCBA) who comes to my home. They give me suggestions on how to decrease the behaviors. After what happened to my brothers and some of my patients, I cannot stand the thought of medicating her.

## Busing

The city farms out its school busing to private agencies that do not always hire reliable, qualified individuals. A student once bit Talia on the bus. Thanks to her winter jacket, her skin was not penetrated. I had to protest with the New York City Office of Pupil Transportation to have a change in her route to prevent this from reoccurring.

Bus drivers and bus aides can be quite lax, and no degree or prior experience with disabled children is required. Companies are not even legally obligated to report abuse, maltreatment, or neglect. Once Talia was brought to school, and the school staff had to bang on the windows to wake up both the sleeping bus driver and aide. In that instance, the bus route was given to a new driver and aide.

I have had aides allow my daughter to eat her lunch on the bus and simply not even deliver it to the school and then try to cover it up. Talia is cognitively unable to remember to deliver her lunch, and I always give it to the bus aide.

Another time, the start date for busing on the letter from the DOE's Office of Pupil Transportation was two business days *before* school started. I immediately called the Office of Pupil Transportation and the bus company, and the driver called me three business days before school began to let me know what time she would be coming the next day. Of course, I told her the school would still not be open. One time a school bus came when the school was closed; I informed the driver the school was closed.

At one placement I visited, I stayed the whole day, as my mother had suggested. Of course, it was not like the 1970s. In my brothers' time, parents were allowed to stay at the potential placement and observe classes all day. In fact, at one school my mother visited, she requested to observe additional classrooms in the afternoon after already watching some classes in the morning. In the afternoon, all the classes she observed were missing a paraprofessional. Now when I visit schools, the staff tell me to leave after only a short time. They cite "confidentiality," although there really is no violation of confidentiality as long no one tells you any student's unique identifying information, such as first and last names, birthdates, or Social Security numbers.

After being outside much of this particular visit, I watched the buses leave. There was a cameraman filming and a reporter from ABC's *Eyewitness News*. I asked the cameraman what he was doing. He told me that a child had been not taken off the bus the previous morning but instead was left on the bus all day and unsupervised. When the bus driver and aide boarded the bus for the afternoon run, they realized what had happened. They then dropped the child off at the babysitter's home as if nothing happened. I was interviewed for this news report. I said this could have been my child. Unfortunately, there have been other cases like this throughout the United States, and some drivers and aides have been criminally charged.

To save money, the DOE's Office of Pupil Transportation has often created bus routes with too many children going to different schools, so that children cannot possibly get to school on time. This has happened to Talia. The DOE will create a new route, but it takes time. For most of us, if we were paying for school ourselves (or any other service), we would want to be on time and get our full value, but the DOE only takes its own budget under consideration.

## Hurricane Sandy: Living with Autism without Electricity

We lost our electricity for five days after Hurricane Sandy, and the electric company said it would be seven to ten days before it would be restored. At first

Talia was agitated, and she went around the house trying to turn on lights. She would occasionally hit herself in the face from this agitation.

We lit candles for light and the stove for heat, but she would try to place her hand over the candles or on the stove. When I knew I could not stay in a room where I lit a candle, I had to blow it out. My home therapists were working with flashlights my husband bought.

The home ABA supervisor suggested I call the electric company to explain our situation and to see if we could receive priority service. I did, and the representative on the phone placed an expedited request, but we still had to wait for two more days.

The homes across from us never lost their electricity. I took Talia and her service dog to ask the neighbors if we could charge Talia's iPad so she could communicate. I offered to pay for the electricity. We were going to use an outside outlet and extend several extension cords approximately 250 feet. The first two refused. One woman told me that she would not let us into her home because she did not know who we were and said we needed to ask "someone else." She still refused even after I explained that we did not need to go into her house but just needed a backyard outlet. I cannot imagine how she thought I could rob her house accompanied by a child with severe autism and a service dog. The third family helped us, and my husband and I later went over with a gift. My husband told her how other people would not help us, and the wife said she was ashamed of them. By the fifth day without power, Talia was calm, sitting on the couch playing with her iPad in the dark room.

## Medicaid Waiver Adventures

Obtaining a Medicaid waiver for Talia took two and a half years. This ultimately gave Talia residential habilitation services that had allowed a provider to come to our home to work with her. It also provided her medical insurance and reimbursement for certain expenses such as her iPad and communication application.

First, we needed to have a psychologist test Talia's IQ. The problem with IQ testing and autism is that the results do not necessarily show progress or identify goals, and the responses are inconsistent. If the score drops, a school district could argue at a hearing that the program was not showing progress, and then the parents could be left holding the tuition bag. By law, parents are required to submit evaluations to their school districts. The psychological testing needed to be done privately, because an Assessment of Basic Language and Learning Skills can be added to the psychological records to show progress. Talia's test was done in December 2009.

To be eligible for Medicaid in New York State, the child has to be receiving a Medicaid service. The first agency I called had no available providers for residential habilitation during Talia's available hours. One issue was that there were few hours available at agencies due to cutbacks. I spent hours on the phone trying to find another agency funded by Medicaid that would submit the application (which takes hours to complete) and begin the residential habilitation. Some agencies never bothered to even return my phone call.

After many months, I found an agency. The application went forward, but the state did nothing for several more months. Talia was turned down by the state because, it said, the psychological test results expired after one year, and this was past December 2010. This was clearly an excuse to turn people down. For people with autism, after the age of five, the IQ remains generally stable, and Talia's IQ was not even close to borderline functioning.

Another excuse was that a physical exam had to occur within six months of the state checking the application, even though physicians perform physical exams annually unless there is a clinical reason to have one more frequently. Private insurance will not pay for an exam more than once a year.

My husband and I also had to disclose all of our financial information. This information is irrelevant and gets waived when applying for services for a child with a disability. After all of this, we finally found providers, though it did not work out well.

The first residential habilitation provider quit after one day because she said that her work hours had been changed. Although she was pleasant and worked as a behavior specialist at a residence, when Talia was hitting herself in the face, she kindly and softly told her, "Don't hurt yourself," which was ineffective. The second provider quit after two days, reportedly because of a family emergency. The third provider lasted about three weeks; she also had a personal reason for leaving.

The fourth provider had physical problems, who sat on my couch while my husband assisted Talia eating her lunch. She was sent to me despite the fact that my daughter was a "runner." My husband took Talia for a walk, and the woman was a block behind. I called the woman that evening and told her not to come back.

The fifth provider could not get to my house. The first week she walked down the wrong block. The next two weeks she did not come, claiming the subways were not running, as if in New York City there was only one mode of public transportation.

The final provider turned out to be a good match for Talia, because she tried to work and play with her. After starting to work with Talia, she decided she wanted to become a special-education teacher and a board-certified behavior analyst.

In September 2012, we received Talia's Medicaid waiver. We could not continue with our current agency residential habilitation providers because it was only for Medicaid-pending applicants. The hunt for another agency began.

The agency that did the initial Medicaid waiver application insisted that, to keep the waiver, we had to enroll Talia in their after-school program. According to the agency representative, this was a Medicaid waiver service. The representative told me there were waiting lists for residential habilitation waiver services, and it would be hard to find an agency with openings. I explained, as I did to this same woman when I initially requested a waiver, that Talia already received therapies after school and that I could not enroll her in that agency's after-school program. She then nastily asked, "Why do you need a Medicaid waiver?"

A few minutes after getting off the phone, she called me back in a sweet voice to tell me about the agency's gala and to ask, since I was a doctor, if perhaps I would like to buy an advertisement. She knew that to get government money, she had to fill the seats in the after-school program, which is why she pressed me to register Talia and threatened that I could lose the Medicaid. I knew this was not true.

I quickly contacted other agencies. I found a community habilitation (formerly called residential habilitation) provider and received authorization for twelve hours of home community habilitation services, although it took two months to get authorization from the Office for People with Developmental Disabilities (OPWDD). The agency and I called the OPWDD to inquire about the problem in receiving authorization. Two weeks after I left a message, someone finally called back. This person left a voice mail telling me that Talia would lose her Medicaid because she was not receiving any waiver services (for which the agency had no authorization to receive payment from the OPWDD). I called back and said that I had left a message two weeks prior. She admitted that she had been out of the office and never checked her messages.

Medicaid selected a managed care health plan, and I needed to find an in-network primary care physician. However, I could not because somehow Talia had been listed as the head of the household, and therefore, I could not speak for her as a legal guardian. I was told to call the Human Resources Administration, which I did. I was on hold for thirty minutes before giving up and leaving my phone number and a request to be called back. Of course, no one called me back.

Two weeks later, I tried again. This time, I was on hold for forty minutes before someone answered the phone. She was not sure how to correct the problem, so she put me on hold for about ten minutes to consult with someone else. There was no music or "thanks for holding" recording, so I had no idea if I had been disconnected. She finally got back on the phone. She would place a complaint with Medicaid, but she was not allowed to give me the complaint number when I requested it. She told me someone would call me back in two weeks, but no one ever did.

Talia's Medicaid service coordinator told me I had to bring Talia down to a Medicaid office to get this changed. Frustrated, I told her that made no sense. According to the law, all ten-year-olds are minors, and their parents are their legal guardians, and I should not have to spend a day at the Medicaid office. The Medicaid service coordinator then called Medicaid again, but she received the same answer. She finally spoke to someone who was able to list me as Talia's legal guardian.

One time, I went to fill a prescription but was told that the Medicaid health plan card had expired. Paying with my credit card, I told the pharmacist that now I knew what my patients went through. I later found out that the insurance card expired if unused within a certain amount of time, but no notice was ever sent to explain this.

A social worker at my job explained that Talia was eligible for straight, rather than managed, Medicaid, but that she would need a new code in the OPWDD system. Straight Medicaid is managed directly by New York State, while managed Medicaid is managed by a private corporation, and there are fewer services available, for example, no long-term nursing visits or transportation to appointments.

This was something my daughter's Medicaid service coordinator should have known. Nevertheless, I told the service coordinator, who received information from Medicaid that I needed a letter from Talia's doctor as to why she needed straight Medicaid. This was incorrect according the social worker at my work. I explained this to the Medicaid service coordinator's assistant. Two days later, someone at the Medicaid office changed Talia's insurance to straight Medicaid. In a complicated labyrinth of useless rules, regulations, and bureaucracy, the people who work for Medicaid do not know what they are doing.

According to OPWDD, families can receive overnight respite for a few days, so parents can take a much-needed vacation or just have a little time for themselves. Since my father became very ill and later passed away, my mother cannot take care of Talia by herself. In 2015, I requested overnight respite in my home to assist my mother care for Talia. Due to Talia's pica, aggression, and elopement, with her need for a service dog, she cannot go to an established agency respite residence. However, my Medicaid service coordinator, after searching for agencies with openings for in-home overnight respite, informed me OPWDD would not fund openings for new cases to receive in-home overnight respite. Consequently, my husband and I have been unable to take vacations together.

Even for people with permanent conditions, like Talia, Medicaid has to be renewed annually. Sometimes, even though the renewal is filed in a timely manner, Medicaid staff can sit on the paperwork for months, and people temporarily lose their Medicaid. That means they cannot have covered medical visits or prescription (and even some nonprescription) medications without privately paying, so they may go without needed medical care.

## Autism Insurance Law

Autism Speaks, in New York as well as other parts of the country, have been proponents of having private health insurance cover autism services such as applied behavior analysis. The insurance helps some individuals. However, just as with other illnesses, the insurance that covers autism in New York requires deductibles and copayments.[52] If people cannot afford their copayments and deductibles for the occasional visit to the internist or gynecologist, they cannot possibly find the funds for daily or almost-daily autism therapies.

In Talia's case, my private insurance is United Healthcare/Oxford. Because of Talia's severe impairments, she needs experienced therapists in order to make progress. Her therapists will not accept the low insurance reimbursement rates, which are lower than the Early Intervention program rates for experienced therapists, and there is no out- of-network option. We also have a seventy dollar copayment per day (until we reach $6,600 a year of out-of-pocket expenses), and ABA requires repetition and consistency over several days each week to be effective. Furthermore, if only private insurance is funding home ABA services, that means individuals without private insurance have

52 "Co-pays and deductibles act as a barrier to care and reduce utilization. The net result is a cost savings (at least in the short run) to the insurance company. In all cases, to a greater or lesser degree, depending on the financial status of the person involved-the deductibles and co-pays discourage or limit utilization. While this save money for the insurance companies, it can be terrible for the patient, especially if that patient is low-income and sick. Co-pays, as we have seen at our community health center…often lead people to delay or completely forego needed care." Bob LeBow, *Health Care Meltdown Confronting the Myths and Fixing Our Failing System* (Boise, Idaho: JRI Press, 2002), 111.

no access to services. Talia's Medicaid does not pay for any ABA under the current autism insurance law.

## Blood Work

When Talia was twelve, she had to have more blood testing done. She was at risk for diabetes, plus her vitamin levels had to be checked because she took mineral oil for constipation, and mineral oil can deplete vitamins. Years ago, her pediatrician did blood work in the office, but insurance does not pay for that anymore. The year before, her blood work was done at Cohen Children's Hospital, which had a restraining chair. Talia does not understand getting a needle and will fight as much as possible. I planned to return to Cohen Children's, but I was told they would only take their own patients, even though Talia's pediatrician was on staff there. I called some labs, but they could not restrain her.

I called New York Presbyterian Hospital in Queens and was relieved when the woman on the phone informed me they could do her blood work with a restraint. When we arrived, there was no restraint. I had to flag down a staff member in the waiting room to watch Talia's service dog because there was not enough space in the phlebotomy room. After waiting for about ninety minutes for all three staff to finish with lunch, I restrained Talia while two staff members held her arm, and another one drew the blood. I nervously sang the A-B-C song, desperately hoping to get all her tests done. It was only because I exercise, run, and lift weights that I was able to hold her. With a more careful diet, her blood sugar improved significantly.

## The Alternative Treatments I Tried

I also tried some alternative therapies-gluten-, casein-, soy-, and corn-free diets-for two months. I also tried vitamin A, fish oil, and even injecting Talia with vitamin B-12. I was like some people in the bargaining stage of a terminal illness. I thought maybe if I tried something, the problem will go away.

I went out of my mind researching and trying to think of recipes on the restrictive diet that also conformed to diabetic modifications for my husband. My daughter did not like eggs or chicken, and I was concerned she was not getting enough protein. I read one study showing children with autism on gluten- and casein-free diets had low tryptophan (a protein-building block) in their blood,[53] which they need to make a chemical called serotonin (a messenger between nerve cells), which has already been found to be lacking in parts of the brain of boys with autism.[54] I tried using vegetarian combinations, such as beans and rice, to make complete proteins, but my daughter did not like that either. Buying the special food for the diet was also emptying my pocketbook. Later in her life, with behavioral therapy, she has now learned to eat just about anything.

Talia became terrified of needles from the B-12 injections. Once, she had a fever, and when she saw the rectal thermometer, she cried so bitterly. I was sure she thought it was a needle. After that, I was done with vitamin B-12 injections.

After these treatments, I did not see improvement. There is insufficient evidence on the gluten- and casein-free diet and fish oil, and the treatments did not help my child. Children may even do better off the gluten- and casein-free diet, because there is more availability for reinforcers in the ABA program. These diets are also socially limiting. It is hard to take a child to a birthday party when they cannot eat the pizza or cake. What makes me angry is the "autism-cure salesman" who distorts the information, gives false hope, and even provides misinformation, so parents cannot make informed decisions regarding treatment for their children.

---

53 Georgianne L. Arnold, Susan L. Hyman, Robert A. Mooney, and Russell S. Kirby, "Plasma Amino Acids Profiles in Children with Autism: Potential Risk of Nutritional Deficiencies," *Journal of Autism and Developmental Disorders* 33, no. 4 (August 2003): 449-54.

54 Diane C. Chugani, Otto Muzik, Robert Rothermel, Michael Behen, Pulak Chakraborty, Thomas Mangner, Ednea A. da Silva, et al., "Altered Serotonin Synthesis in the Dentatothalamocortical Pathway in Autistic Boys," *Annals Neurology* 42, no. 4 (October 1997): 666-69, doi:10.1002/ana.410420420.

## Accepting Talia's Autism and Finding Joy

It was hard for me to accept my child would not be cured, and it took me a few years to accept it. I also wished she was just higher functioning, until I started treating some "high-functioning" autistic children who were not functioning well at all and caused a lot of grief to their parents.

However, after accepting Talia's autism, I made peace with myself. I realized that when I stopped wishing she did not have autism, I could love her fully the way she was. Of course, I still want her to do the best that she can, and I want to provide any opportunity possible to facilitate that just like I do with my other child.

And, from time to time, I still think about what my life would be like if I did not have an autistic child-how I would not have to worry about money and about the freedom and relaxation I would have. I think about the nice trips I would take, like I used to. I recall the memories of the beauty of riding a mule to the bottom of the Grand Canyon; walking through the bamboo forest of Nara, Japan; strolling down the beach in Israel with my husband; and enjoying a cocktail by the pool with no worries on my mind. I think of my younger days as a student in Thailand, when I ran down the street during the New Year celebration, trying to escape the young men chasing me with buckets of water to dump over my head, according to their custom. They always ran faster than me, and, except for my ruined camera and water-damaged passport, it was a carefree experience. However, my situation is what it is, and I have to deal with it.

I also think autism is something for which we should seek a cure. Despite what some high-functioning people with autism believe, as far as I am concerned, when someone cannot communicate hunger, thirst, a need to use the bathroom, or even pain, that person needs treatment, just as with any illness that impairs functioning. I also know the people who do not want a cure will not be up at night worrying and are not going to care for my child when I am dead.

One week short of Talia's eleventh birthday, we were out for a walk, and we passed a park. Talia turned to me and said, "Park." It was the first time she had spontaneously done that.

I joyfully responded, "Sure, we can go to the park." Talia laughed on the swing, and it was another reminder that all these years of advocating for her (and my brothers too), I had done the right thing.

From diagnosis to the years of struggles for treatment that will never end, it gives my life meaning to see Talia's progress. If the society would support rather than create obstacles to treatment for its weakest people and their families, we and our children's symptoms, functioning, and quality of life could be so much better. The rewards are priceless.

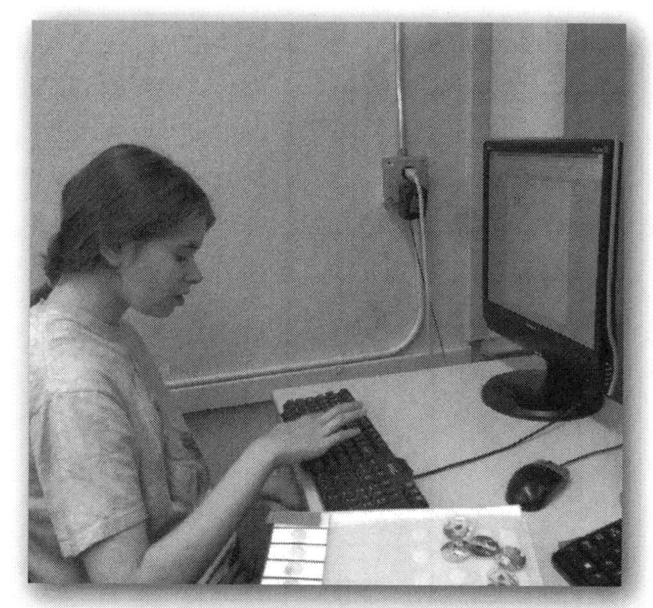

Talia typing her name at school.

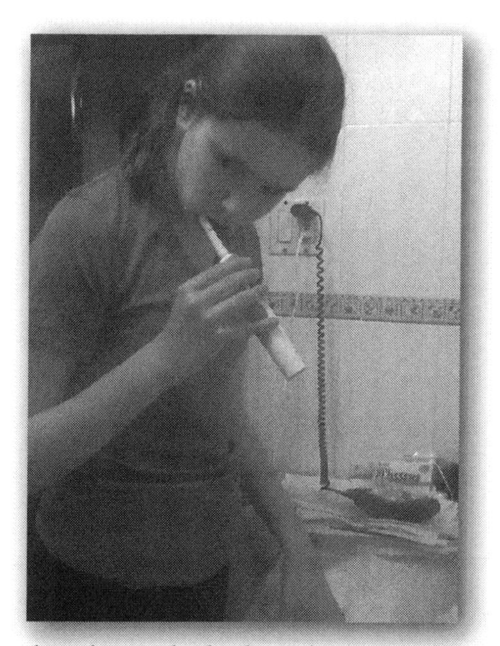

Talia brushing her teeth; this has taken years of hand-over-hand repetition, broken down into small steps.

Talia outside with her service dog, Cecil. She holds the handle independently, and the dog stops her from running.

Talia enjoys being a DJ on a school trip.

**Water activities are Talia's favorite.**

# Five

## Research and Clinical Applications-Positive-Only Treatments, Aversives, Medications, Developmental Disabilities, and Advocacy

This chapter discusses the research, politics, and clinical applications of aversive shock treatment, positive-only treatments, and medications used with the developmentally disabled population. It discusses the legal ramifications and how special interests promote distortions of research, which result in poor-quality, nonexistent, and even dangerous care. This chapter also discusses advocacy for the right to have research evidence-based treatment.

### Shock Controversy

The JRC is controversial for its use of behavioral skin shocks administered to the leg or arm, an aversive procedure I mentioned earlier. I have tried the skin-shock device. The only side effect is temporary skin redness or perhaps some anxiety between the time of the behavior and the skin-shock application, which is never longer than two minutes, and the majority are delivered within one minute.

A group of people who believe in "positive-behavior supports" condemn this therapy. Eric Schopler, the originator of the TEACCH program at the University of North Carolina, described them perfectly. In one article, "He calls the critics 'self-serving ideologues' who drastically over simplify the issue with emotional arguments and 'are making a fortune going around doing workshops on how to never use aversives.'"[55] They also get to sell their books and do their consultations, and it does not matter to them if people who do not respond to their therapies get killed in the process. Their own research, in an analysis of 109 articles, shows the frequency of problem behavior can be suppressed by 90 percent in only about half of subjects using their positive approaches alone,[56] which is not sufficient for a life-threatening behavior, especially when suppression needs to be at 100 percent.

By contrast supplementary aversives are by far much more successful in research to treat life-threatening behavior.[57] In one published study conducted at the JRC with sixty individuals, "[w]hen end-of-baseline data were compared with end-of-treatment data, CSS [contingent skin-shock], as a supplement to positive programming, showed effectiveness (defined as a 90% or greater reduction from baseline) with 100% of the participants...Psychotropic medications were reduced by 98%, emergency takedown restraints were reduced by 100%, and aggression-caused staff injuries were reduced by 96%. As a result of the treatment, 38% of participants no longer required CSS and some returned to a normal living pattern."[58] Improvements with the skin-shock device in some of the students could be seen in the graphs below.[59]

---

55 Constance Holden, "What's Holding Up 'Aversives' Report," Science 249, no. 4972 (August 31, 1990): 981.

56 Edward G. Carr, Robert H. Horner, Ann P. Turnbull, Janet G Marquis, Darlene Magito McLaughlin, Michelle L. McAtee, Christopher E. Smith, et al., *Positive Behavior Support for People with Developmental Disabilities: A Research Synthesis* (Washington, DC: American Association on Mental Retardation, 1999), 45.

57 For a discussion of some of these articles, see Dorothea C. Lerman and Christina M. Vorndran, "On the Status of Knowledge for Using Punishment: Implications for Treating Behavior Disorders," *Journal of Applied Behavior Analysis* 35, no. 4 (2002): 431-64, doi:10.1901/jaba.2002.35-431.

58 Matthew L. Israel, Nathan A. Blenkush, Robert E. von Heyn, and Patricia M. Rivera, "Treatment of Aggression with Behavioral Programming that Includes Supplementary.

59 Israel, Blenkush, von Heyn, and Rivera, "Treatment of Aggression with Behavioral Programming that Includes Supplementary Contingent Skin-Shock," 128.

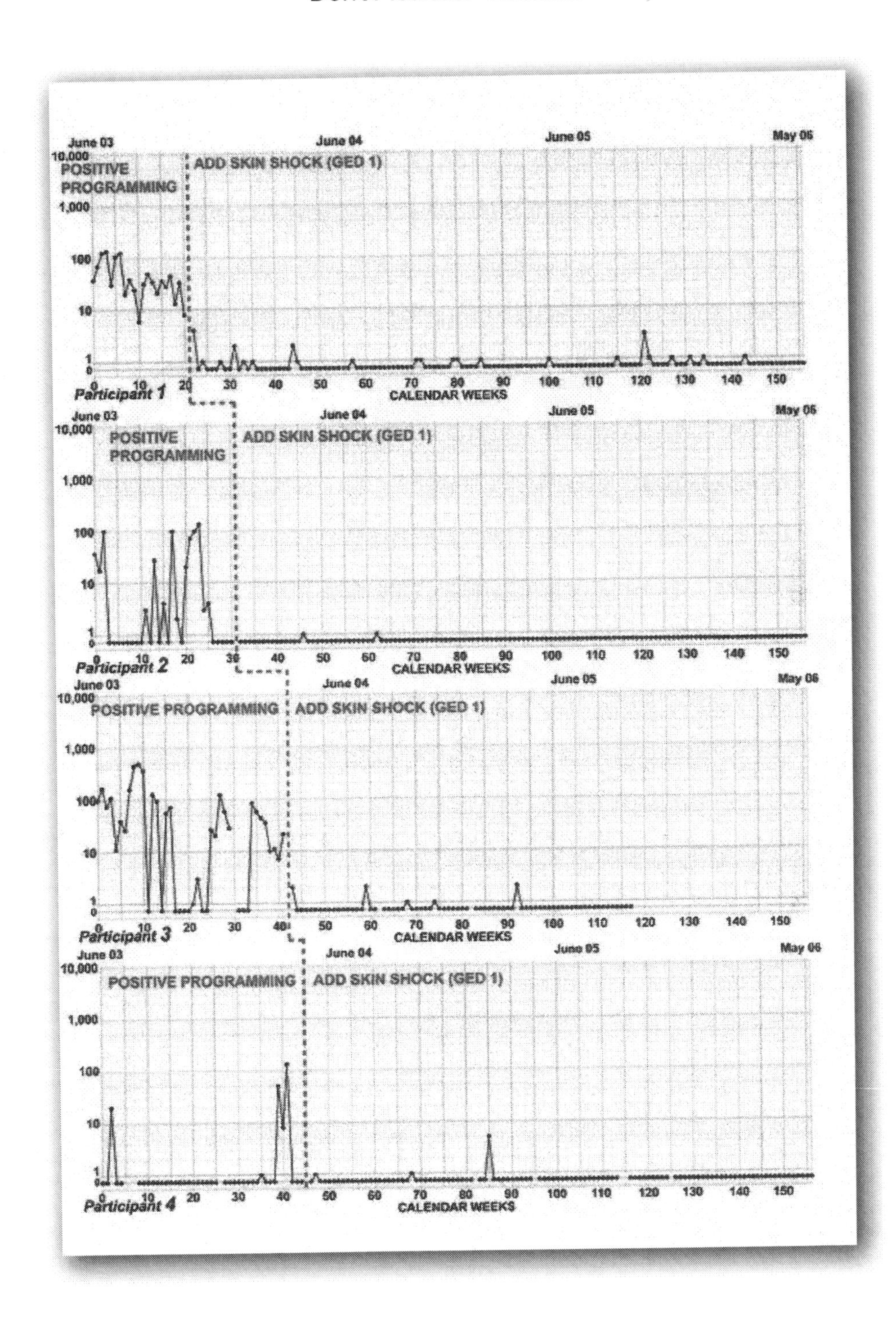
June 03
June 04
June 05
May 06
10,000
1,000
100
10
1
0
POSITIVE PROGRAMMING
ADD SKIN SHOCK (GED 1)
Participant 1
CALENDAR WEEKS
0 10 20 30 40 50 60 70 80 90 100 110 120 130 140 150
Participant 2
Participant 3
Participant 4

Another study has also shown that using aversives stops the need for physical restraints,[60] which has been and remains deadly. According to one article, an eleven-year-old boy died while being restrained in a psychiatric hospital.[61] Another article described how a sixteen-year-old boy went into cardiac arrest and subsequently died at a residential treatment center during physical restraint: "No criminal charges are warranted in the death of Corey Foster, a 16-year-old boy who went into cardiac arrest while being physically restrained at a residential treatment center in Yonkers, the Westchester County district attorney concluded Thursday after a three-month investigation."[62]

Just as at other times in history, the anti-aversive movement uses pseudoscience, half truths, and even false statements and emotional arguments to defend their position. As discussed earlier, 119 peer-reviewed articles support the efficacy of skin-shock treatment for behaviors.[63] Furthermore, the research on positive-only treatments shows it has significant limitations, including defining "severe" behaviors as behaviors that are not life threatening and would never warrant aversive treatment.[64] Yet even if the behavior is dangerous, and

60 Pieter C. Duker and Daniel M. Seys, "A Quasi-Experimental Study on the Effect of Electrical Aversion Treatment on Imposed Mechanical Restraint for Severe Self-Injurious Behavior," *Research in Developmental Disabilities* 21, no. 4 (July-August 2000): 235-42, doi:10.1016/S0891-4222(00)00039-1.

61 Dwight F. Blint and Colin Poitras, "Boy, 11, Crushed during Restraint," *Hartford Courant*, March 24, 1998, http://articles.courant.com/1998-03-24/news/9803240179_1_boy-s-death-dcf-officials-medical-examiner-s-office.

62 Nina Bernstein, "No Charges in Death at Yonkers Center for Youths," *New York Times*, August 2, 2012, http://www.nytimes.com/2012/08/03/nyregion/no-charges-to-be-filed-in-death-of-corey-foster-at-a-yonkers-treatment-center.html.

63 Bibliography on Skin-Shock, accessed January 24, 2016, http://www.effectivetreatment.org.

64 The positive-behavior support movement has defined behaviors as severe that are not really severe. For example, in a book-Edward G. Carr, Len Levin, Gene McConnachie, Jane I. Carlson, Duane C. Kemp, and Christopher E. Smith, *Communication-Based Intervention for Problem Behavior: A User's Guide for Producing Positive Change* (Baltimore, MD: Brookes, 1994)-and later in a chapter titled "Hypothesis-Based Intervention for Severe Problem Behavior"-Edward G. Carr, Nancy A. Langdon, and Scott C. Yarbrough, *Functional Analysis of Problem Behavior*, ed. Alan C. Repp and Robert H. Horner (Pacific Grove, CA: Wadsworth, 1999), 9-31-the behaviors discussed were "screaming, spitting at the teacher, grabbing another child's hair or throwing her school work off the desk,'" R. M. Foxx, "Severe Aggressive and Self-Destructive Behavior: The Myth of the Nonaversive Treatment of Severe Behavior," in *Controversial Therapies for Developmental Disabilities*, 304-5. Nobody will die from these behaviors.

the individual does not respond to the positive-only techniques, these "advocates" will be against any aversive intervention. What happens when an individual is dangerous is that the individual will be referred to a psychiatrist for medication management, which is not always effective and has side effects. Unlike aversives, which are only used after positive-behavior interventions are proven to fail, medications will be prescribed right away for a dangerous behavior. Apparently these "advocates" and the collaborative politicians have decided that it is better that a person gets side effects from medications such as diabetes, heart disease, obesity (which can cause up to eight years of loss of life and nineteen years of loss of health from type 2 diabetes and cardiovascular disease),[65] cancer, bone loss, tardive dyskinesia, seizures, neuroleptic malignant syndrome, liver failure, or other adverse effects than receive a two-second skin shock, on average, less than once a week or any other aversive.

It might seem ridiculous to object to a two-second skin shock once a week to prevent all the above illnesses. Yet others who have no training in medicine, who have never even met our children, and who have never walked a mile in our shoes are choosing for them all those illnesses, over a two-second skin shock. How many people with cancer or diabetes, if told they could be free from their illness completely with a two-second skin shock once a week, would say no?

Individuals with autism can have sensory impairment.[66] In *Diagnostic and Statistical Manual of Mental Disorders* (fifth edition), one of the diagnostic criteria (although not necessary to be present) for autism-spectrum disorder is "[h]yper- or hyporeactivity to sensory input or unusual interest in sensory aspects of the environment (e.g. apparent indifference to pain/temperature, adverse response to specific sounds or textures, excessive smelling or touching

---

65 Science Daily, "Obesity May Shorten Life Expectancy by up to 8 Years and Cut Healthy Life by up to 19 years," December 5, 2014, https://www.sciencedaily.com/releases/2014/12/141205094841.htm;
Steven A Grover, Mohammed Kaouache, Philip Rempel, Lawrence Joseph, Martin Dawes, David C. W. Lau, and Ilka Lowensteyn, "Years of Life Lost and Healthy Life-Years Lost from Diabetes and Cardiovascular Disease in Overweight and Obese People: a Modelling Study," *Lancet Diabetes & Endocrinology* 3, no. 2 (February 2015): 114-22, doi:10.1016/S2213-8587(14)70229-3.

66 Fry Williams and Lee Williams, *Effective Programs for Treating Autism Spectrum Disorder: Applied Behavior Analysis Models,* 12.

of objects, visual fascination with lights or movement)."[67] For example, a noise that most individuals have no problem with can be very aversive to an autistic person. On the flip side, banging one's head would cause severe pain that most others would avoid, but it may not be painful at all or may even be positively reinforcing to someone with autism.

While most people with autism do not and will not require the use of skin shocks, and no two people with autism are the same, aversive skin shock compensates for that impairment of not feeling pain appropriately. The inability to feel pain or other uncomfortable aversive sensations can be deadly. For example, if we did not feel thirsty or hungry, we would die of dehydration or starvation. If we did not feel cold, we would not mind going out in a bathing suit to play in the snow, and we may die from exposure. Without pain, people may burn parts of their bodies before realizing anything is wrong.

Pure positive-behavior approaches create "positiveland" for individuals with dangerous behaviors, like the witch's house in *Hansel and Gretel.* On the outside, it looks like candy, and everything is positive. The staff in such a facility can only respond to individuals in a positive fashion. Unlike for typical children and adults, there are no negative consequences for behaviors. People have negative consequences even for not coining parking meters, and I doubt people would coin parking meters just for some positive reinforcement, like the meter playing a little music or having a brightly colored message with some verbal praise, such as "Thank you. You help make New York strong!" or some tangible reinforcer, such as the parking meter dispensing a piece of candy.

However, unlike in our world, in the world of individuals with developmental disabilities, subjected to positive-only approaches, nothing can be taken away. Of course, no aversives are allowed. What can be wrong with that? The problem is that, instead of an effective behavior plan, these facilities use toxic medications to manage dangerous behavior. These medications can kill people, like the witch's oven in *Hansel and Gretel.*

By contrast, aversives research shows that using skin shock actually produces positive side effects.

---

67 *Diagnostic and Statistical Manual of Mental Disorders*, 5th ed. (Arlington, VA: American Psychiatric Publishing, 2013), 50.

One study stated this:

> The rate of self-injurious head hitting was reduced using contingent electric shock delivered via the Self-Injurious Behavior Inhibiting System (SIBIS). Positive side effects indicating an improved affective state and increased interaction with the environment were documented. Treatment gains were maintained at a 1-year follow-up assessment. The consistent reports of positive affective side effects from successful treatment studies using SIBIS and contingent electric shock are noted…The behaviors, Laugh, Smile, "Doggie," Self-Initiated Toy Play, and Self-Stimulation were felt to be indicative of positive side effects because they suggested either a positive affective state or increased interaction with the environment. All of these behaviors increased from baseline levels during treatment with SIBIS.[68]

Another study, this one at the JRC, found the following:

> [T]he side effects of contingent shock (CS) treatment were addressed with a group of nine individuals, who showed severe forms of self-injurious behavior (SIB) and aggressive behavior. Side effects were assigned to one of the following four behavior categories: (a) positive verbal and nonverbal utterances (b) negative verbal and nonverbal utterances (c) socially appropriate behaviors, and (d) time off work. When treatment was compared to baseline measures, results showed that with all behavior categories, individuals either significantly improved, or did not show any change. Negative side effects failed to be found in this study.[69]

---

68 Thomas R. Linscheid, Carrie Pejeau, Sheila Cohen, and Marianna Footo-Lenz, "Positive Side Effects in the Treatment of SIB Using the Self-Injurious Behavior Inhibiting System (SIBIS): Implications for Operant and Biochemical Explanations of SIB," *Research in Developmental Disabilities* 15, no. 1 (January-February 1994): 81, 85.

69 W. M. W. J. van Oorsouw, M. L. Israel, R. E. von Heyn, P. C. Duker, "Side Effects of Contingent Shock Treatment." *Research in Developmental Disabilities* 29, no. 6 (November-December 2008): 513, doi:10.1016/j.ridd.2007.08.005.

Four other studies on skin shock have also found positive side effects.[70]

There are those who buy into "THE FALSE MANTRA that 'America has the best health care system in the world,'"[71] until they get sick and are saddled with health-care bills that can drive people into bankruptcy. In the same fashion, it is easy to believe there is no need to use aversives until you have a loved one with life-threatening behaviors who has developed serious life-threatening adverse effects from psychotropic medications that are not even effective in treating the dangerous behaviors.

It is true, as opponents often claim, that aversives can be abused. However, so can other treatments, including psychiatric medication. It would be irresponsible to ban either treatment, as each can help people function and can save lives. Treatments need to be individualized for people. To ban aversives to protect people from abuse, even though the therapy saves lives, would be sacrificing lives to benefit a group. When a society decides to sacrifice lives to benefit a group, we all become at risk.

The worst victims of the anti-aversive movement are those individuals with disabilities who are being neglected and abused, physically and sexually, at schools, adult day programs, and residences, and who cannot communicate their neglect and abuse, despite existing laws. They are receiving underfunded, poorly implemented, inconsistent behavior plans and are not learning (and may even be regressing) at schools, day programs, and residences. They

---

70 Thomas R. Linscheid and Heidi Reichenbach, "Multiple Factors in the Long-Term Effectiveness of Contingent Electric Shock Treatment for Self-injurious Behavior: a Case Example," *Research in Developmental Disabilities* 23, no. 2 (March-April 2002): 161-77, doi:10.1016/S0891-4222(02)00093-8; Sarah-Jeanne Salvy, James A. Mulick, Eric Butter, Rita Kahng Bartlett, and Thomas R. Linscheid, "Contingent Electric Shock (SIBIS) and a Conditioned Punisher Eliminate Severe Head Banging in a Preschool Child," *Behavioral Interventions* 19, no. 2 (April 2004): 59-72, doi:10.1002/bin.157; Thomas R. Linscheid, Brian A. Iwata, Robert W. Ricketts, Don E. Williams, and James C. Griffin, "Clinical Evaluation of the Self-Injurious Behavior Inhibiting System (SIBIS)," *Journal of Applied Behavior Analysis* 23, no. 1 (Spring 1990), 53-78; R. M. Foxx, R. G. Bittle, and G. D. Faw, "A Maintenance Strategy for Discontinuing Aversive Procedures: A 52-Month Follow-Up of the Treatment of Aggression," *American Journal on Mental Retardation* 94, no. 1 (1989): 27-36.

71 LeBow, *Health Care Meltdown Confronting the Myths and Fixing Our Failing System*, 14.

are being bullied by peers, including housemates and roommates.[72] They are harming themselves and others. They are medicated with toxic drugs, sedated into living vegetables, and living with life-threatening side effects. They are getting injured during physical restraints. These are victims for whom the advocacy groups maintain silence while nonstop pursuing their anti-aversive and anti-JRC agendas. People who claim that aversives are torture, even though they are a thoroughly proven and researched form of therapy, obviously do not need deadly medications to control their children and are fortunate compared to the group of severely handicapped people. If the government can deny one treatment (even if it is the only effective treatment) to one person for one condition, then the government can deny any treatment to any person for any condition, even if there is no other effective treatment.

## Ineffective and Nonexistent Behavior Plans

By federal law, all school-age individuals who have a behavior that interferes with learning require a functional behavior assessment and a behavior intervention plan. Sometimes, the behaviors all get lumped together (as was done with Talia's IEP), with one intervention for all behaviors, although different behaviors can have different functions and will need different intervention plans. At one school that has referred me patients with severe behaviors, students had "verbal calming and redirection" on the behavior plan, regardless of whether or not the behavior was attention-seeking in nature, and whether or not the verbal response could increase the frequency of the behavior.

Once when I was in a foreign country, I had an upset stomach and could not keep anything down. A clinic gave me some yellow pills. A professor with

---

72 Medicaid regulations mandate "individuals having choice regarding roommate selection within a residential setting," *Fact Sheet: Summary of Key Provisions of the Home and Community-Based Services (HCBS) Settings Final Rule (CMS 2249-F/2296-F)*, Department of Health & Human Services Centers for Medicare & Medicaid Services, January 10, 2014, http://www.medicaid.gov/medicaid-chip-program-information/by-topics/long-term-services-and-supports/home-and-community-based-services/downloads/hcbs-setting-fact-sheet.pdf. However, due to a lack of available residential placements, individuals often have no alternative setting to relocate.

a sore throat told me the clinic gave him what appeared to be the same yellow pills. This school just gave all the students "the same yellow pills."

Some students never have a functional behavior assessment or get a behavior plan, even when their behavior interferes with learning. In my experience, almost all schools and agencies for older individuals with developmental disabilities in New York are not sufficiently trained to do functional experimental behavior analysis in which different conditions are manipulated to see how they affect the frequency of a behavior and to determine the likely cause(s) of the behavior.

Functional behavior assessments and behavior intervention plans may still not be done, even if a school calls 911 because of the behavior, even when the child is on numerous psychotropic medications to control that behavior. When 911 is called, the police, who are generally untrained to work with special-needs individuals, arrive. Sometimes they treat the disabled individual like a criminal. The person is handcuffed and threatened with and sprayed with pepper spray, which subsequently traumatizes the individual. As stated earlier, police took Stuart away from his group home in handcuffs for his behavior.

In the hospital, the person may be placed into physical restraints or be chemically sedated. Of course, we do not use the term "chemical restraint." The professionals may refer to the behavior as manic, psychotic agitation, or whatever else, but the result is the same. If behavioral interventions are not available, and the doctor does not prescribe medications when someone is dangerous, it is malpractice.

Schools will call parents to hurry to pick their children up to avoid having the school call 911. This may occur more than once a week. Day programs may drop off the individual at home or have a family member or residence staff member pick up the person. A residence staff member once informed me that her supervisor instructed the staff not to pick up the phone every time it rings to avoid picking up people from day programs because of their behaviors. As I will discuss in greater detail later in this book, schools will call the Administration for Children's Services (ACS) to accuse parents of medical neglect when the parents refuse to have their children medicated. Anti-aversive organizations have nothing to say about any of that.

Calling 911 can negatively or positively reinforce behavior. The individual gets to avoid doing work and gets lots of attention. For some people, physical restraint for being dangerous is positively reinforcing. Calls to 911 and physical restraints can seriously increase the frequency of dangerous behavior.

Some of my cognitively impaired patients have been taken advantage of and even physically and sexually abused by general psychiatry patients in psychiatric hospitals. As mentioned earlier, Stuart was neglected in a hospital. Psychiatric hospitals generally do not want to admit people with severe developmental disabilities, especially if they require a 1:1 sitter. Many hospitals are not equipped to treat and have nothing to offer other than sedating medication, because other therapies-such as group psychotherapies or other therapeutic activities, including group relaxation or ceramics-are not appropriate for many individuals with significant intellectual disabilities.

At one OPWDD-funded agency with which I work, residents with intellectual disabilities were allowed to refuse to see the psychologist who would develop a behavior plan. This meant someone could be without a behavior plan, even if the person was on multiple medications with toxic side effects to control dangerous behavior.

Group homes and day programs collect behavior-frequency data, but often the data are inaccurate because there are missing days of data collection or inaccurate dates. When data have to be turned in at the end of the month, staff may simply write what they can recall. One time, a staff member wrote data up to November 30, 2012, even though it was only November 26.

When I was a day-habilitation consultant, I sometimes asked the staff about behavior, and they would tell me about problems occurring days before, yet there was no documentation. The staff would say they did not write it down yet. I have had patients for whom all that was submitted prior to an appointment was a combined composite total frequency of all behaviors, ranging from teasing to yelling and to physical aggression. Teasing is not a symptom of a psychiatric condition that can respond to medication, whereas physical aggression can be. Sometimes these poorly collected data will make it to the psychiatry appointments, and sometimes they will not. There is no standard in New York State for separating topographies of behavior.

Instead of seriously monitoring abuse (which I discuss later) with cameras, open-door policies, or research evidence to improve behavior plans, the OPWDD relies on unscientific political correctness such as Person Centered

Planning. This is how person-centered approaches (PCAs) work: "PCAs envisage a new, positive, inclusive future for the focus person," which Dr. Beth Mount considers superior to "traditional forms of planning" that focus on "the deficits and needs of people, overwhelming people with endless program goals and objectives, and assigning responsibility for decision-making to professionals."[73]

Dr. Grayson Osborne has noted that the PCA committees are formed by family, friends, and focus members rather than professionals. The committee discusses "what they know of the focus person's wishes, capacities, vision of the future, and if necessary, challenging behavior…Unfortunately, the opinions of others may not validly map the preferences of individuals with profound disabilities."[74] Regarding PCAs, he has also noted, "the goals they do contain are vague and not easily measured,"[75] and that without measurement, there is no accountability.[76] Furthermore, without measurements, there can be no scientific research to establish effectiveness.

I was stunned at Talia's OPWDD Individualized Service Plan (ISP), which stated (in the spirit of PCAs) in a section titled "Valued Outcomes"-they do not use the word "goals": "Talia would like to increase her ability to live independently by performing tasks with minimal assistance. Talia would like to improve her social skills as exhibited by her ability to establish social relationships. Talia would like to increase her community integration. Talia would like to communicate effectively with her family, peers and supervisors."

I told the Medicaid service coordinator, who wrote the plan, that what Talia would really like to do was eat cookies all day. She explained that if she changed it, she did not know if she would get into trouble. There was nothing

---

73 Beth Mount, "Benefits and Limitations of Personal Futures Planning," in *Creating Individual Supports for People with Developmental Disabilities: A Mandate for Change at Many Levels*, ed. Valerie J. Bradley, John W. Ashbaugh, and Bruce C. Blaney (Baltimore, MD: Brookes, 1994), 98.

74 J. Grayson Osborne, "Person Centered Planning: A Faux Fixe in the Service of Humanism?" in *Controversial Therapies for Developmental Disabilities: Fad, Fashion, and Science in Professional Practice*, ed. John W. Jacobson, Richard M. Foxx, and James A. Mulick (Mahwah, NJ: Lawrence Erlbaum Associates, 2005), 317.

75 Ibid., 319.

76 Ibid., 314.

measurable in these valued outcomes, so there was no accountability. In 2012, "person-centered" approaches were still politically correct in New York State, despite their lack of accountability. The 2012 OPWDD regulations included "Person-Centered Behavioral Intervention."[77] The regulations, of course, not only banned the use of aversives but also forbade removing individuals' "rights" to cell phones or entertainment electronic devices for disciplinary reasons.[78] The federal government also mandates person-centered planning.[79]

---

77 CRR-NY 633.16NY-CRR: Official Compilation of Codes, Rules and Regulations of the State of New York, title 14, Department of Mental Hygiene Chapter XIV, *Office for People with Developmental Disabilities*, part 633, Protection of Individuals Receiving Services in Facilities Operated and/or Certified by OMRDD, https://govt.westlaw.com/nycrr/Document/Ieb7ffe02978311e29e9f0000845b8d3e?viewType=FullText&originationContext=documenttoc&transitionType=CategoryPageItem&contextData=(sc.Default).

78 *Rights limitations*: "The limitation of a person's rights as specified in section 633.4 of this part (including but not limited to access to mail, telephone, visitation, personal property, electronic communication devices (e.g., cell phones, stationary or portable electronic communication or entertainment devices computers), program activities and/or equipment, items commonly used by members of a household, travel to/in the community, privacy, or personal allowance to manage challenging behavior) shall be in conformance with the following:...[R]ights shall not be limited for the convenience of staff, as a threat, as a means of retribution, for disciplinary purposes or as a substitute for treatment or supervision." CRR-NY 633.16NY-CRR: Official Compilation of Codes, Rules and Regulations of the State of New York, title 14, Department of Mental Hygiene Chapter XIV, *Office for People with Developmental Disabilities*, part 633, Protection of Individuals Receiving Services in Facilities Operated and/or Certified by OMRDD, j2, https://govt.westlaw.com/nycrr/Document/Ieb7ffe02978311e29e9f0000845b8d3e?viewType=FullText&originationContext=documenttoc&transitionType=CategoryPageItem&contextData=(sc.Default);
'Time-out is a restrictive/intrusive intervention in which a person is temporarily removed from reinforcement or denied the opportunity to obtain reinforcement *and* during which the person is under constant visual and auditory contact and supervision. Time-out interventions include:
(a) placing a person in a specific time-out room, commonly referred to as exclusionary time-out;
(b) removing the positively reinforcing environment from the individual, commonly referred to as non-exclusionary time-out." Official Compilation of Codes, Rules and Regulations of the State of New York, title 14, Department of Mental Hygiene Chapter XIV, Office for People with Developmental Disabilities, part 633, Protection of Individuals Receiving Services in Facilities Operated and/or Certified by OMRDD, j3, https://govt.westlaw.com/nycrr/Document/Ieb7ffe02978311e29e9f0000845b8d3e?viewType=FullText&originationContext=documenttoc&transitionType=CategoryPageItem&contextData=(sc.Default).

79 "Home and Community Based Services," *Centers for Medicare & Medicaid Services*, January 10, 2014,
https://www.cms.gov/Newsroom/MediaReleaseDatabase/Fact-sheets/2014-Fact-sheets-items/2014-01-10-2.html.

This shows the anti-aversive movement is so out of control and that individuals over the age of twenty-one living at residences and attending day programs in New York State cannot have "time outs" or have reinforcing items withheld as a consequence, such as their CD players (which can be purchased with an individual's state-given allowance) for problem behavior-interventions parents typically do with children-even if the behavior is dangerous or life threatening. Even if the preferred item is given to the individual by the legal guardian, and the legal guardian requests removing the item as a consequence for a behavior, the staff cannot do so.

Withholding an item is considered a human rights abuse. The reinforcing items can only be withheld in an emergency. With no other option to manage behavior, I submit, some individuals are medicated instead and potentially get sick or die from the side effects. The state has unilaterally decided it is better that an individual is medicated-which may cause the adverse effect of diabetes, cancer, or another major medical illness-than have a material item temporarily removed.

For parents, it is considered acceptable to withhold television or take away a cell phone for a limited time period, if a child does not behave. This benign intervention, which teaches a lesson, is preferable to placing children on high doses of psychotropic medication. Yet for our population of developmental disabled over the age of twenty-one, the state chooses the medication.

Most OPWDD-funded facilities that I have worked with, even those with a population who have dangerous behaviors, do not hire board-certified behavior analysts (BCBAs). The residences will hire what they call "applied behavioral science specialists," or more recently, "behavior intervention specialists," who do not need to have any licensing, take any board exams, or work a specific number of hours under supervision,[80] unlike a board-certified

---

80 Definition Behavioral Intervention Specialist, 14 CRR-NY 633.16NY-CRR: Official Compilation of Codes, Rules and Regulations of the State of New York, title 14, Department of Mental Hygiene Chapter XIV, *Office for People with Developmental Disabilities*, part 633, Protection of Individuals Receiving Services in Facilities Operated and/or Certified by OMRDD, b32, https://govt.westlaw.com/nycrr/Document/Ieb7ffe02978311e29e9f0000845b8d3e?viewType=FullText&originationContext=documenttoc&transitionType=CategoryPageItem&contextData=(sc.Default).

behavior analyst, who must pass a formal certification process that includes working a certain number of hours under supervision and passing an exam.[81] The psychologists at the residences and day programs often do not have the knowledge or experience to develop effective behavior plans.

One residence staff member expressed his frustration about a typical PCA policy: when an individual threw an electronic device and broke it, the agency just bought that person another one. At the same residence, the consequence for a severe behavior was the staff asking the individual what help the individual wanted. In other words, it taught the patients to be aggressive to get something. While I understand as the human rights access to food, shelter, and medical care, entertainment devices can hardly be considered "rights."

Staffers informed me that in 2015, OPWDD residences had to stop using handicapped parking passes for individuals with developmental disabilities with the exception of physical impairments. The handicapped parking passes were considered stigmatizing. However, what really is stigmatizing is when someone is exhibiting dangerous behavior in the community due to the lack of an effective behavior plan.

Some group homes and day programs in New York State allow individuals, even nonverbal individuals, to eat anything they want. I submit the state considers it a human right for individuals at residences to routinely purchase their junk food with their own allowances (provided with state money) or earnings. They may consume at the residence items, such as a half box of cereal or a container of juice at once. It is not a human right to allow someone who cannot make an informed decision to destroy him- or herself. If someone is actively suicidal, it is malpractice to allow the person to commit suicide. However, these individuals are allowed to kill themselves in the name of "human rights," with its excessive choices, just as long as the process is slow and unintentional. To compensate for their meager budgets, day programs sell junk food, so individuals with disabilities have opportunities to finance their care with their health, just like many typical schoolchildren finance their

81 Behavior Analyst Certification Board, Experience Standards, accessed September 4, 2016, http://bacb.com/wp-content/uploads/2015/08/150824-experience-standards-english.pdf.

education with vending machines. The person-centered approach supports choices, even if they can be disastrous.[82]

Many regulations promote "human rights," but do not assert a right to a research-validated functional behavior assessment, including a functional experimental analysis, if needed to compose an effective plan, rather than using potentially harmful medication. This is something the regulations have nothing to say about. Being so limited in functional behavior assessments and behavior plans makes individuals untrainable. Individualized Service Plans use the term "person centered" despite the fact that person-centered planning lacks accountability.

The regulations specify that at day programs individuals on psychotropic medications who only have diagnoses such as impulse control disorder or autism and do not receive medication for solely a co-occurring psychiatric diagnosis such as schizophrenia must have behavior plans.[83] However, when further clarified by the OPWDD, this is only if the medication is administered at the program, although drugs remain in the body after administration.

Another frustration I find is that New York State residential programs do not keep records on why previous medications were discontinued. It is sometimes impossible to find out about all the prior medication trials of a patient. This is detrimental to patient care.

Many day programs for individuals over the age of twenty-one have no windows in the classrooms, despite medical and psychological benefits of natural light.[84] In one report it was stated, "Daylighting has been associated with

82 With PCAs, individuals can make harmful choices, such as quitting their job or making poor dietary choices, Osborne, *Controversial Therapies for Developmental Disabilities: Fad, Fashion, and Science in Professional Practice*, 320-21. Individuals have such choices even though they may not understand the consequences of their decisions. Therefore, PCAs can interfere with health and structure, and individuals may not develop their full potential.

83 NYCRR 633.16NY-CRR Official Compilation of Codes, Rules and Regulations of the State of New York, title 14, Department of Mental Hygiene Chapter XIV, *Office for People with Developmental Disabilities*, part 633, Protection of Individuals Receiving Services in Facilities Operated and/or Certified by OMRDD, c8, j5, https://govt.westlaw.com/nycrr/Document/Ieb7ffe02978311e29e9f0000845b8d3e?viewType=FullText&originationContext=documenttoc&transitionType=CategoryPageItem&contextData=(sc.Default).

84 Norman E. Rosenthal, *Winter Blues, Everything You Need to Know to Beat Seasonal Affective Disorder* (New York, NY: Guilford Press, 2006), 5.

improved mood, enhanced morale, lower fatigue, and reduced eyestrain… [N]atural views…may be more effective in reducing stress, decreasing anxiety, holding attention, and improving mood."[85]

The report also states in its conclusion: "[N]atural light has proved to be beneficial for the health, productivity, and safety of building occupants. Natural light helps maintain good health and can cure some medical ailments. The pleasant environment created by natural light decreases stress levels for office workers."[86] Therefore, the lack of sunlight in the classrooms of individuals with disabilities may cause anxiety and mood disturbances resulting in further agitation.

## Lack of Programs

Some adults are home with family caregivers, because day programs cannot manage their behaviors. When the participants have behaviors the day programs cannot manage, they will be suspended or terminated, and if the participants are living at home, they will be sent right back to be cared for by an overwhelmed caregiver, sometimes an elderly parent who is in no position physically or emotionally to deal with an unsafe family member. If the day program, with their professionals, cannot handle the behavior, it seems unreasonable to expect an elderly parent to manage a dangerous problem. Sometimes the day program will suspend the individual and will not allow the individual to return until the individual has been on psychotropic medication for weeks. Many parents have had to leave their jobs to take care of family members in these circumstances.

An individual with an unmanageable behavior from a state-run group home is much harder to suspend or discharge from a day program than an individual living at another type of home. The state will claim it is the person's human right to attend a day program, even though it should never

85 L. Edwards and P. Torcellini, *A Literature Review on the Effects of Natural Light on Building Occupants* (Golden, CO: National Renewable Energy Laboratory, July 2002) NREL/TP-550-30769, 4, http://www.nrel.gov/docs/fy02osti/30769.pdf.

86 Edwards and Torcellini, *A Literature Review on the Effects of Natural Light on Building Occupants*, 38.

be a human right to attend an inappropriate placement. Sometimes, group homes will threaten to expel residents if parents refuse to consent to potentially deadly medications to manage behavior, when behavior plans are so limited and practiced so inconsistently, due to poor training, supervision, and inflexible regulations that prohibit potentially effective procedures. It is not so different than at Willowbrook (a state institution), when, to gain admission for their children, some parents had to consent to receive the hepatitis infection for experimental purposes. "For example, when the main school was closed to new admissions in 1964 due to overcrowding, parents were told there were openings in the hepatitis unit for children who could participate in the study. The public outcry over this case was largely due to the impression that parents had little choice over whether or not to participate in the research. Parents who wanted care for their children may not have had any other options."[87] I have also observed some consent forms that group home staff composed that omitted some of the potential life-threatening side effects. Some individuals with dangerous behaviors will not be accepted into a residence at all.

In programs for school-age children, this reliance on medication would be illegal, as medications are not allowed to be part of a child's Individualized Education Program (IEP). A child cannot be forced to be without an education, whereas people over the age of twenty-one have no such protections. New York State will also use the excuse of behavior problems to deny access to home health aides, creating further stress on elderly parents.

In general psychiatry, when someone shows behavioral symptoms, and general outpatient treatment is not effective, the individual is referred to a higher level of care, such as intensive outpatient treatment, partial (day) hospital, or inpatient care. Yet with the developmentally disabled, when their behaviors escalate at a day program, the level of care in most cases is decreased; that is, the individual is simply sent home, as happened to my brother Stuart. Unlike with education for those under the age of twenty-one, those who are

87 "Willowbrook Hepatitis Experiments," in *Exploring Bioethics* (Waltham, MA: Education Development Center, 2009) 4, https://science.education.nih.gov/supplements/nih9/bioethics/guide/pdf/master_5-4.pdf.

older have no right to appropriate treatment, due process with impartial hearings, or right to appeal to obtain what they need, except, at least in New York, in the case of transferring residential placements or changing a service. In New York, even those hearings are heard by a three-member panel made of direct employees of OPWDD, although it is possible, but costly, to appeal to a federal judge. Therefore people over the age of twenty-one do not have rights to effective services.

Unlike in IEPs, where specific staffing ratios are written down and have to be enforced according to federal law, the Individuals with Disabilities Education Improvement Act (IDEIA), in placements for those over the age of twenty-one, if there are budget cuts, staffing is simply cut at the day program or residence, which can result in more behavior problems, because the individuals are not getting their specific needs met. It becomes impossible to properly implement a behavior plan or even collect data effectively. Furthermore, as there is no federal law for individuals over the age of twenty-one that mandates staff credentials, the instructors at adult day programs, unlike at schools, have no specific special education or developmental disabilities certification requirements, and sometimes there may only be one person with only a high school diploma supervising the class, even if there are supposed to be three staff members in the room.

The regulations for care for those individuals over the age of twenty-one are so poor, in adult day program where I consulted, that the staff could stay on their cell phones or text for much of the day without consequences, which is unacceptable in a school setting. No one supervised some of the individuals when they used the restroom to make sure they were not touching anything unsanitary or washing their hands. As has occurred with some individuals I have worked with, instead of a 6:1:1 ratio that they had at school, mandated by their IEPs, the adults were placed in a class of fifteen with sometimes only one supervising adult present. If the IEP for someone under the age of twenty-one provided for a 1:1 crisis paraprofessional, that individual most likely will lose that paraprofessional when entering a program for those over the age of twenty-one as has occurred with some of my patients since the IDEIA no longer applies.

I have some patients who attended schools-public as well as private-prior to the age of twenty-one. Some of them were in 6:1:1 classes. As with Stuart, some required medications or increased amounts of medications after entering their adult day programs. I have even discharged patients because residences told me they would not implement an intense behavior plan that I have recommended-such as a differential reinforcement for other behavior at five-minute intervals to treat a serious behavior problem-because the residence refused to provide the 1:1 staffing necessary to implement the plan while the individual is heavily medicated. My plan would provide a reinforcer every five minutes for not engaging in a target behavior. At one New York State residence, the staff wrote that my patient did not need the 1:1 staffing I requested, even though their own psychologist told me, "Deep down, I agree with you." There was nothing that could be done.

Even with limited budgets, much of it is paid to top executives, with some earning close to $1 million a year, along with perks such as a luxury company car. This has occurred under the OPWDD's watch.

Because there is no right to a hearing, unlike with the municipal school districts that participate in IEP meetings, the OPWDD does not participate in Individualized Service Plan (ISP) meetings unless the individual resides at a state facility. School districts have to fund all the IEP services, but with ISPs, the state hands out limited budgets to private agencies. The agencies that cannot afford necessary services conduct the meetings for the individuals, so the individual is short changed, while the state washes its hands clean of the whole process. There is no one present who controls funding to whom one can advocate for a needed service. A representative of those in charge of the purse strings is nowhere in sight.

These budgets are already limited, especially because New York State often audits the agencies that serve individuals with disabilities, and when documentation is missing in one sampled file, the state will assume other individuals' files are also missing documentation, and the state will fine the agencies. What this amounts to is that innocent individuals with disabilities are penalized rather than the individual who failed to document the files properly.

However, what is not audited until it is too late is administrative overspending and corruption-funds not going to the individuals with the disabilities. According to a *New York Daily News* article, "One of the city's largest nonprofits has suddenly discovered it is $19.4 million in the red and plans to make major cutbacks to its social service programs."[88]

The following article from the *New York Times* found similar problems:

> Medicaid money created quite a nice life for the Levy brothers from Flatbush, Brooklyn.
>
> The brothers, Philip and Joel, earned close to $1 million a year each as the two top executives running a Medicaid-financed nonprofit organization serving the developmentally disabled.
>
> They each had luxury cars paid for with public money. And when their children went to college, they could pass on the tuition bills to their nonprofit group.
>
> Philip H. Levy went as far as charging the organization $50,400 for his daughter's living expenses one year when she attended graduate school at New York University. That money paid not for a dorm room, but rather it helped her buy a co-op apartment in Greenwich Village.[89]

The lack of funding results in larger class sizes, less residential staff, a lack of crisis paraprofessionals, and unqualified staff. This, in turn, results in increased psychotropic medication use, which is in my opinion unethical. Also, by farming out the care to private entities with shoestring budgets, when something does go wrong, the state can negate responsibility because the incident occurred at a private agency.

---

88 Reuven Blau, "Federation Employment & Guidance Service Has $19.4 Million Shortfall," *New York Daily News*, December 12, 2014, http://www.nydailynews.com/new-york/fegs-nyc-non-profit-disabled-19-4m-shortfall-article-1.2043351.

89 Ross Buettner, "Abused and Used Reaping Millions in Nonprofit Care for the Disabled," *New York Times*, August 2, 2011, http://www.nytimes.com/2011/08/02/nyregion/for-executives-at-group-homes-generous-pay-and-little-oversight.html.

Furthermore, in April 2016, OPWDD stopped funding speech and language and other therapies in residences, although improved communication can decrease problem behaviors, as discussed elsewhere. OPWDD has told therapists that they will need to join a group to continue to provide these services but created no such "group" at the time when they terminated funding.

Furthermore, although the necessary care is underfunded, unnecessary things, even sometimes things with problematic consequences, do get funded. For example, the OPWDD funds free cell phones for individuals with developmental disabilities, even though they sometimes used them to call people in the middle of the night or tie up 911. Some disabled individuals require constant supervision, so there is no need for them to have cell phones.

I have a patient under the age of twenty-one for whom I requested a 1:1 crisis paraprofessional, and the school psychologist told me she herself recommended it but was told to remove the recommendation. She told me this in confidence, but that school staff were not allowed to recommend 1:1 paraprofessionals. However, in this situation (unlike if the person were over the age of twenty-one), at least the parent could request a hearing, and the funding came directly from the school district. For busing, sometimes an individual gets agitated with rides that take more than two hours one way getting to a program. On an IEP, a student can have limited-time travel, and students under the age of twenty-one in New York City never ride more than two hours. However, for those over twenty-one, there are no time limits on bus rides.

I have patients who are verbal but totally unproductive. They are allowed to refuse to go to any day program and just sit around and watch television all day. Indeed the federal regulations support this by mandating "[t]he individual controls his/her own schedule including access to food at any time."[90] If you or I wanted to stay home and watch television all day, we would not be able to pay for our food or shelter. If we allowed our children to refuse

90 *Fact Sheet: Summary of Key Provisions of the Home and Community-Based Services (HCBS) Settings Final Rule (CMS 2249-F/2296-F),* Department of Health & Human Services Centers for Medicare & Medicaid Services, January 10, 2014, https://www.medicaid.gov/Medicaid-CHIP-Program-Information/By-Topics/Long-Term-Services-and-Supports/Home-and-Community-Based-Services/Downloads/HCBS-setting-fact-sheet.pdf.

school and let them watch television all day, the authorities would remove them from our care. However, these residents live in a concocted world where doing nothing is an option. Many of them do not exercise; stay up too late with pleasurable activities; eat poorly; do not get enough sleep (which further increases agitation and possibly causing obesity or diabetes);[91] and refuse medical appointments, resulting in, among other problems, that their psychotropic medications are not being monitored properly, and they do not understand the medical consequences of their choices. Therefore, they are not making informed decisions.

## "Age-Appropriate" Activities Are Inappropriate

My mother's friend, who had a special-needs child, would say, "He's twenty-one going on five." In residences and day programs in New York State, individuals must have "age-appropriate" activities, rather than activities appropriate to their individual cognitive and developmental level. This generalized approach of "age appropriate" further limits the few activities they can participate in, given their cognitive limitations.

At a program where I did consulting, a staff member, who worked in a room with individuals over the age of twenty-one who had profound intellectual disabilities complicated by deafness or blindness, told me how state officials came in and forced the staff to remove some items they considered "not age appropriate," such as building blocks. Items such as crayons and wooden puzzles were forbidden too. A lack of stimulation will increase behavior problems and the need for medication management.

A developmentally disabled woman I was treating always brought a doll to the appointment. She held this doll with affection and called it her "baby." I was dismayed one day when a staff member informed me that the doll was taken away because it was considered not age appropriate. I told the staff that having that doll was the closest she would ever be to having her own baby, and

91 Kristen L. Knutson, Karine Spiegel, Plamen Peney, and Eve Van Cauter, "The Metabolic Consequences of Sleep Deprivation," *Sleep Medicine Reviews* 11, no. 3 (June 2007): 163-78, doi:10.1016/j.smrv.2007.01.002.

this was her way to express her maternal instincts. She ultimately was allowed to keep the doll.

The forbidden age for inappropriate activities can be decided arbitrarily. In one group home, the manager would not permit a trip to see *The Lion King* on Broadway, citing it was not age appropriate. I find it ironic that the above items are forbidden activities, while television or music cannot be forbidden as a way to manage dangerous behavior. The decisions to allow only "age appropriate" activities or to ban withholding preferred items were made on the state level, and they only serve to increase psychotropic medication use.

## Shock and the Media

The media likes to sensationalize. A CBS report narrated by Connie Chung about the JRC edited the sounds of the skin-shock device to make it longer than two seconds. One critique stated the following:

> Elsewhere, scenes of the shock device at work, apparently provided by the institute, were broadcast in slow motion to a spooky sound. In one of the more original excuses of the television era, CBS News blamed the institute for the producers' use of these gimmicks. A CBS spokeswoman told the Boston Globe that the institute did not give the producers enough footage to fill the available time, so they had to "stretch what we had."[92]

In other media coverage, Geraldo Rivera, without ever visiting the program, compared the treatment to how dogs were cared for. *Primetime* interviewed me but never used any of the footage for their show.

*Mother Jones* magazine produced a negative report on Matthew's school. To get my story and the stories of others, the reporter, Jennifer Gonnerman,

---

92 Walter Goodman, "Television View; a Few Scary Pictures Can Go a Long Way," *New York Times*, March 20, 1994, http://www.nytimes.com/1994/03/20/arts/television-view-a-few-scary-pictures-can-go-a-long-way.html?pagewanted=all; David Armstrong, "School Says CBS Altered Sound in Report on Its Shock Therapy," *Boston Globe*, March 9, 1994.

misidentified herself as a reporter for the *New York Times.* She also did not use my interview for her story, because she already knew her conclusion.

One argument often made against the JRC is the cost. Admittedly, the cost of educating and housing these individuals is high wherever they are. *ProPublica* interviewed me for an hour for an article,[93] but never used anything I said. The author mentioned in the title that New York City spent $30 million a year to send students there. However, I would like to see a control cost computed. For example, the cost of multiple medications can be thousands of dollars each month. In addition, the JRC saves money by not using 911 for behavioral emergencies but handles the behaviors themselves, rather than by using far more costly psychiatric hospitalizations. Psychiatric hospitalizations will cost much more than the JRC per day. Also, the costs of the medical complications of the medications, such as diabetes, heart failure, or seizures would be even more. Also, some students in other residential programs require 1:1 or even 2:1 staffing. These combined expenses can easily cost more than the care provided at the JRC. The *ProPublica* article also cited Dr. Gregory Fritz, a child psychiatrist who stated that aversives can "lead to trauma."

Regarding the *ProPublica* article, I responded to Eric Umansky, an assistant managing editor of *ProPublica*:

> I am writing concerning the recent article by Heather Vogell and Annie Waldman (http://www.propublica.org/article/nyc-sends-30-million-a-year-to-school-with-history-of-giving-kids-shocks). Annie Waldman interviewed me for one hour by phone for this article. I have identical twin brothers and a daughter who all have autism. One brother resides at the Judge Rotenberg Center (JRC), and one resides at a facility in New York. I am also a practicing psychiatrist who completed a fellowship at the Seaver and New York Autism Center at Mount Sinai Hospital.
>
> As I explained to Ms. Waldman, my brother at JRC has been there for almost twenty-six years. Prior to being at JRC, he banged

93 Heather Vogell and Annie Waldman, "New York City Sends $30 Million a Year to School with History of Giving Kids Electric Shocks," *ProPublica,* December 23, 2014, https://www.propublica.org/article/nyc-sends-30-million-a-year-to-school-with-history-of-giving-kids-shocks.

his head into sharp objects requiring surgery. Since arriving at JRC, [he] has been doing well with no medication and has not needed an aversive for over two years.

Contrast this to his identical twin brother, who resides in New York and has no access to aversives. He has repeatedly been in the hospital, once for over five months, and is so heavily medicated that I have called him after two in the afternoon only to find he was still sleeping. He is only forty-three years old and currently takes twenty medications, most of which are for his behaviors and to treat the side effects of the psychiatric medication. I can think of no better relevant situation to compare the effectiveness of treatments between JRC and in-state facilities for some patients than to compare twin brothers, as well as to explain to your readers why some parents and guardians choose to send their family members to JRC rather than an in-state facility. Yet, Ms. Waldman did not include this in her story for what I assume was its failure to fit into her preconceived bias against JRC.

Ms. Waldman also informed me that she had read my published article on aversive treatment in Psych Congress Network (http://www.psychcongress.com/blogs/ilana-slaff-galatan-md/skin-shock-saved-my-brother%E2%80%99s-life), where I cited published research articles showing positive benefits from the treatment, such as improved affective state, increased interaction with the environment, and self-initiated toy play and that the gains were maintained at a one year follow-up assessment. (Linscheid TR, Pejeau C, Cohen S, Footo-Lenz M. Positive side effects in the treatment of SIB using the self-injurious behavior inhibiting system (SIBIS): Implications for operant and biochemical explanations of SIB. Res Dev Disabil 1994;15(1):81-90.

Yet Ms. Waldman chose not to mention anything about my statements, which conflicted with her agenda. She only used the interview of a professional who was against aversives without any interview of an independent professional with an opposing view. Indeed I had informed her that a private expert independent of Judge Rotenberg Center had stated and even documented in a report that my brother who lives at Judge Rotenberg Center would be at risk to his life if he were transferred to the proposed New York placements I was given. I also informed her of a textbook chapter, "Severe Aggressive and Self-Destructive Behavior: The Myth of Nonaversive Treatment of Severe Behaviors" (Jacobson JW, Foxx RM, Mulick JA. Controversial Therapies for Developmental Disabilities. Lawrence Erlbaum Associates; 2005: 304-305), which she told me she would review.

In addition, I informed Ms. Waldman of individuals who came back to New York and died from their behaviors and how currently... former students after returning to...have been at...Hospital for years, many times the price of residing at the Judge Rotenberg Center. The Judge Rotenberg Center accepts children who have been rejected for placement at every other facility in the United States, as happened to my brother and still happens to others. Without the Judge Rotenberg Center, would we consider these children hopeless? Do we keep them chemically restrained in a psychiatric hospital setting? If other interventions are effective, how does it happen that no other placement in the country will accept some of these children?

The article was misleading when discussing suspension. Suspension implies that parents are told to go pick up their children. Judge Rotenberg Center never calls parents to pick up their children, but rather if they are a danger to themselves or others, including their classmates, for everyone's safety they need to be placed in an alternative room. The use of suspension should have been honestly clarified.

It is also completely unbalanced and even dishonest to only mention the aversives without the positive-behavior interventions that comprise most of the JRC program. My brother is on seven positive-behavior contracts at once and earns preferred items throughout the day for being safe. I had also informed Ms. Waldman that almost all other placements will not offer that intensive positive-behavior reinforcement.

I would like to know why this information was excluded from Ms. Waldman's and Ms. Vogell's story. It seems that they went out of

> their way to portray JRC in the most negative way possible-so much so that they were blind to the factual information that was provided to them by me and perhaps others who they interviewed for this story.
>
> Thank you for your attention to this matter.
>
> P.S. The same day the article was published the school had to close due to a bomb threat. Inaccurate and incendiary reporting can lead to violence, including the loss of innocent lives.

Here is Mr. Umansky's response:

> Thank you for your letter. I would simply note that while your brother is an adult our story focused on JRC's treatment of school-age children. Also, if you'd like to share your thoughts further, I encourage you to post them as a comment on the story itself.

In addition to being interviewed on *ProPublica*, Dr. Fritz also spoke out against aversives in a CBS report.[94] In the CBS report, it was stated that Dr. Fritz "said there's virtually no reliable data this type of aversive shock therapy is effective in bringing about a change in behavior, over the short or long term, once the shock is withdrawn." Contrary to Dr. Fritz's statement, five published research studies describe individuals with dangerous behaviors who maintained their treatment gains after the shock devices were faded.[95] After the CBS report, I

94 Amy Burkholder and Anna Werner, "Controversy over Shocking People with Autism, Behavioral Disorders," *CBS News*, August 5, 2014, http://www.cbsnews.com/news/controversy-over-shocking-people-with-autism-behavioral-disorders/.

95 Foxx, Bittle, and Faw, "A Maintenance Strategy for Discontinuing Aversive Procedures: A 52-Month Follow-Up of the Treatment of Aggression," 27-36; Thomas R. Linscheid, Fred Hartel, and Nannette Cooley, "Are Aversive Procedures Durable? A Five Year Follow-Up of Three Individuals Treated with Contingent Skin Shock," *Child and Adolescent Mental Health Care* 3, no. 2 (1993): 67-76; Don E. Williams, Sharon Kirkpatrick-Sanchez, and W. Terry Crocker, "A Long-Term Follow-Up of Treatment for Severe Self-Injury," *Research in Developmental Disabilities* 15, no. 6 (1994): 487-501; Salvy, Mulick, Butter, Kahng Bartlett, and Linscheid, "Contingent Electric Shock (SIBIS) and a Conditioned Punisher Eliminate Severe Head Banging in a Preschool Child," 59-72; Israel, Blenkush, von Heyn, and Rivera, "Treatment of Aggression with Behavioral Programming That Includes Supplementary Contingent Skin-Shock," 119-66.

sent Dr. Fritz a link to my published article on skin shock in *Psych Congress*, and I also sent him this reply:

> Good afternoon. I have some concern regarding your statement to CBS that there is no evidence for efficacy in changing behavior once the skin-shock is withdrawn. As psychiatrists, we prescribe antipsychotics and mood stabilizers for individuals with chronic schizophrenia and bipolar disorder. As we know, these medications have serious adverse effects and also in patients with the above conditions there is no evidence of efficacy once those medications are withdrawn. There is no evidence insulin injections will be effective for diabetics once the insulin is stopped. The same applies to countless other treatments, yet we use them to keep individuals alive and functioning. I feel it is unethical to hold the skin-shock to a different standard as the alternative for my brother and others can and will be death. Thank you for reading this.

Not surprisingly, he never even acknowledged me.

Sleepless one night, I thought how I would love to have him do a one-to-one debate with me on aversive skin shock. I doubt he ever read any of the 119 peer-reviewed articles. I started to think how I would have to spend hours preparing, memorizing the studies. Then I realized there was no way any network would air it, and even if they did, Dr. Fritz would most probably never agree to it.

Colleagues like him make me feel disillusioned with my profession. We physicians should focus on evidence-based medicine-but how the profession is confounded with politics and the pharmaceutical industry. I wonder if I am reading scientific journal articles and attending scientific presentations or I am just reading or watching advertisements from colleagues who have received funds from-and therefore may have lost their objectivity or perhaps are even sold out to-the pharmaceutical industry. But then I think of my supportive and caring colleagues, who are so understanding, helpful, and sympathetic. I cannot allow Dr. Fritz or any other colleague to make me feel isolated and turn me away.

## Psychiatric Visits and Unclear Diagnoses

Diagnostic validity is not very precise in developmental disabilities. Hyperactivity may be attributed to attention-deficit hyperactivity disorder rather than attributed to someone with autism. This can happen if the individual with autism never had intensive, properly done ABA to learn how to sit appropriately. This includes identifying reinforcers with properly done assessments, rewarding the behavior of going to an adult, making the adult a conditioned reinforcer, and sitting at a table with initially very frequent reinforcers. These reinforcers may even initially need to be given at intervals of less than a minute. According to one article, "children with ASD [autism spectrum disorder] may mistakenly be diagnosed with ADHD because they have autism-related social impairments rather than problems with attention."[96]

Emotional instability can be diagnosed as bipolar disorder. When a two-year-old throws temper tantrums, we think this is what two-year-olds do. However, someone in their twenties throwing temper tantrums can be diagnosed with bipolar disorder. Someone who is unable to have organized behavior due to profound intellectual disability can be diagnosed with a psychotic disorder. If an individual with autism and poor emotional regulation and lack of social cognition becomes aggressive and needs to be hospitalized for safety, the person may receive a diagnosis of bipolar disorder.

I have had colleagues misattribute fecal smearing as a symptom of mood instability or even psychosis and then recommend increases in medication. Individuals with autism with sensory impairments may find fecal smearing to be positively reinforcing. As stated earlier, my daughter did, and the solution was not medication, but rather to temporarily give her a special cookie every time she had a bowel movement in the toilet.

The staff going to a psychiatry appointment from an OPWDD-funded residence may be simply direct-care staff without even a high school

96 "Children With Autism May Be Over Diagnosed With ADHD," *Neuroscience News*, October 28, 2016, http://neurosciencenews.com/adhd-autism-psychology-5381/; Benjamin E. Yerys, Jenelle Nissley-Tsiopinis, Ashley de Marchena, Marley W. Watkins, Ligia Antezana, Thomas J. Power, and Robert T. Schultz, "Evaluation of the ADHD Rating Scale in Youth with Autism," *Journal of Autism and Developmental Disorders*, published online October 13, 2016, doi:10.1007/s10803-016-2933-z.

diploma and may not even spend much time with the patient but may still feed information to the psychiatrist to make major medication decisions. Sometimes, no written information is sent to the visit, such as behavioral data or a report from a psychologist. I once had a Spanish-speaking-only residential staff member come to an appointment with a patient who had an intellectual disability and limited expressive language skills, and who spoke only English.

### *More Diagnoses Equal More Money*

Psychiatric visits are paid for by Evaluation and Management (E&M) codes with reimbursement from Medicaid and Medicare, depending on the number of diagnoses. Someone with lots of diagnoses can get more payments than someone with few diagnoses, even though the visit is for the same amount of time. Therefore, there is a financial incentive to come up with as many diagnoses as possible. Unfortunately, more diagnoses can mean more medication. Also, some diagnoses have more reimbursement than others.

## Medications: Overuse, Misuse, and Dangers

Psychiatric medications are given extensively to individuals with autism. "Medications of some kind are used in approximately half of young patients with autism in the United States, but should only be applied in cases where benefits outweigh the risks of side effects."[97] "Age data indicate that about 70% of children with autism-spectrum disorders age 8 yr and up receive some form of psychoactive medication in a given year."[98] "Children with ASD are much more prone to the side effects of these medications compared with the general population."[99] "Each year, millions of Americans are hospitalized (and thousands

97 Amar Mehta, "Family Complications in Autism Diagnosis," *Psychiatric Annals* 42, no. 8 (August 2012): 289, doi:10.3928/00485713-20120806-04.

98 Donald P. Oswald and Neil A. Sonenklar, "Medication Use among Children with Autism Spectrum Disorders," *Journal of Child and Adolescent Psychopharmacology* 17, no. 3 (July 2007): 348, doi:10.1089/cap.2006.17303.

99 Azeem, Imran, and Khawaja, "Autism Spectrum Disorder: An Update," 61.

die) due to adverse drug reactions."[100] Pharmaceuticals obtain high profits from medicating individuals with autism. "The global market for autism therapeutics is presently estimated at between $2.2 billion and $3.5 billion."[101] Whether many of these medications are effective is highly questionable. "Many children with autism are prescribed psychotropic drugs, sometimes several at once for long periods of time, according to a new study…Many of these agents [drugs] have yet to be proven effective for treating autism."[102]

The self-appointed anti-aversives advocates never speak out about the risks of medication, other than a general statement about preventing chemical restraint. They do not address off-label prescribing of medication to manage behavior, which is prevalent. Many of the medications used are not approved for children, but these are used on them anyway and are used on some individuals in untested combinations of multiple drugs.

These same self-appointed advocates have nothing to say about medication abuse, when individuals are so medicated that they are falling down and injuring themselves, because the facility had no applied behavior analysis available but the individuals were dangerous. Some drug cocktails consist of seven or more medication combinations. Furthermore, polypharmacy increases the risk for medication administration errors and is particularly difficult for families to keep track of when they take their children out of residential settings for home visits. "[D]rug treatment is prescribed at 'disturbingly high' rates."[103]

---

100 Salma Malik and Muhammad Waqar Azeem, "Psychopharmacogenomics in Pediatric Psychiatry with a Focus on Cytochrome P450 Testing," *Psychiatric Annals* 46, no. 1 (January 2016): 53, doi:10.3928/00485713-20151130-01.

101 Bryan King, "Power to the Placebo: An Update on Pharmacotherapy in Autism," *Mount Sinai School of Medicine Advances in Autism Conference,* April 11, 2010.

102 Kathryn Doyle, "Autistic Kids Often Get Multiple Psychotropic Drugs At Once," *Psych Congress Network*, October 24, 2013, http://www.psychcongress.com/article/autistic-kids-often-get-multiple-psychotropic-drugs-once-13726); Donna Spencer, Jaclyn Marshall, Brady Post, Mahesh Kulakodlu, Craig Newschaffer, Taylor Dennen, Francisca Azocar, et al., "Psychotropic Medication Use and Polypharmacy in Children with Autism Spectrum Disorders," *Pediatrics* 132, no. 5 (November 1, 2013): 833-40, doi:10.1542/peds.2012-3774.

103 Fry Williams and Lee Williams, *Effective Programs for Treating Autism Spectrum Disorder*, 17; National Institutes of Health Consensus Development Panel, "Treatment of Destructive Behaviors in Persons with Developmental Disabilities," *Journal of Autism and Developmental Disorders* 20, no. 3 (1990): 403-29.

## *Pharmaceuticals Pay the FDA to Approve Medications That Have Questionable Evidence*

Medication approvals are regulated by the FDA. "In 1992 Congress passed the Prescription Drug User Fee Act, under which drug companies pay a variety of fees to the FDA, with the aim of speeding up drug approval (thereby making the drug industry a major funder of the agency set up to regulate it)."[104] Despite that the industry has to pay the FDA regardless whether the drug is approved or rejected, this still amounts to a potential conflict of interest, because it creates an inherent financial incentive for the FDA to approve drugs, as if the FDA were to require more research, pharmaceutical companies would be more discerning and submit less drugs for approval until more thorough evidence is established. Therefore, the FDA would lose part of its earnings. The user fees also create another potential conflict of interest, as these payments may remove objectivity when examining the research to make a determination. Furthermore, the FDA does not require the raw data in research to be released to the public for further scrutiny,[105] and industry sponsors the preponderance of controlled studies to establish evidence.[106] In fact, although an initial published study, using data submitted from and funded by SmithKline Beecham-later changed to GlaxoSmithKline, manufacturer of Paxil (paroxetine)-demonstrated effectiveness for Paxil (paroxetine) for adolescent major depression,[107] a reanalysis of the same data showed no significant improvement from placebo and only a significant risk of harm including

104 Carl Elliott, "The Drug Pushers," *Atlantic Monthly,* April 1, 2006, http://www.theatlantic.com/magazine/archive/2006/04/the-drug-pushers/304714/.

105 Debra Kroszner, "Public Health Groups Sue FDA for Disclosure of Clinical Trial Data for Costly Hep C Drugs," *Information Society Project Yale Law School,* June 25, 2015, http://isp.yale.edu/node/5978.

106 Philip G. Janicak and Joseph Esposito, "An Update on the Diagnosis and Treatment of Bipolar Disorder, Part 1: Mania," *Psychiatric Times,* no. 11 (November 20, 2015), http://www.psychiatrictimes.com/cme/update-diagnosis-and-treatment-bipolar-disorder-part-1-mania.

107 Martin B. Keller, Neal D. Ryan, Michael Strober, Rachel G. Klein, Stan P. Kutcher, Boris Birmaher, Owen R. Hagino, et al., "Efficacy of Paroxetine in the Treatment of Adolescent Major Depression: A Randomized, Controlled Trial," *Journal of the American Academy of Child & Adolescent Psychiatry* 40, no. 7 (July 2001): 762-72, doi:10.1097/00004583-200107000-00010.

adverse events such suicidal ideation and behavior.[108] In practice, the FDA's autism medication registration clinical trials have serious flaws.

There are only two FDA-approved medications to treat irritability in autism, Risperdal (risperidone) and Abilify (aripiprazole), both of which are antipsychotics. The FDA approval for Risperdal stated, "The efficacy of RISPERDAL in the treatment of irritability associated with autistic disorder was established in two 8-week, placebo-controlled trials in children and adolescents (aged 5 to 16 years) who met the DSM-IV criteria for autistic disorder," and that the studies "measured the emotional and behavioral symptoms of autism, including aggression towards others, deliberate self-injuriousness, temper tantrums, and quickly changing moods."[109] The FDA approval for Abilify stated, "The efficacy of ABILIFY (aripiprazole) in the treatment of irritability associated with autistic disorder was established in two 8-week, placebo-controlled trials in pediatric patients (6 to 17 years of age) who met the DSM-IV criteria for autistic disorder and demonstrated behaviors such as tantrums, aggression, self-injurious behavior, or a combination of these problems."[110]

However, when the data in the studies used to obtain the FDA approval are closely analyzed, there is no clear evidence these medications help with self-injurious behavior at all. The studies all used the Aberrant Behavior Checklist, Irritability Subscale (ABC-I) as outcome measures in placebo-controlled trials. The ABC-I has fifteen items, some of which do not involve physical behaviors. Only three involve self-injury. The ABC-I also does not measure exact frequencies of behaviors (which is the gold standard of measurement in behavior analytic research) but rather depends on caregivers' recollection, which

108 "Reanalysis of JAACAP Study on Paroxetine Sparks Controversy," *American Psychiatric Association Psychiatric News Alert*, September 17, 2015, http://alert.psychnews.org/2015/09/reanalysis-of-jaacap-study-on.html; Joanna Le Noury, John M. Nardo, David Healy, Jon Jureidini, Melissa Raven, Catalin Tufanaru, and Elia Abi-Jaoude, "Restoring Study 329: Efficacy and Harms of Paroxetine and Imipramine in Treatment of Major Depression in Adolescence," *British Medical Journal* 351, h4320 (September 16, 2015), doi:10.1136/bmj.h4320.

109 "Risperdal US Full Prescribing Information," Janssen Pharmaceuticals, Titusville, NJ, revised April 2014, 14.4, http://www.janssenpharmaceuticalsinc.com/assets/risperdal.pdf.

110 "Abilify US Full Prescribing Information," Otsuka Pharmaceutical, Tokyo, Japan, revised January 2016, 14.4, http://www.otsuka-us.com/Documents/Abilify.PI.pdf.

is considerably less precise. The FDA registration trials of risperidone did not publish the breakdown of the individual items.[111]

In addition, Johnson and Johnson, the manufacturer of Risperdal, agreed to pay $2.2 billion in criminal and civil fines for inappropriately marketing the drug to individuals with developmental disabilities in addition to other populations.[112] The marketing occurred prior to the FDA approval for Risperdal for autism.

Furthermore, in placebo-controlled trials of Abilify (aripiprazole), clinical trials that received funding from the pharmaceutical company, Bristol Myers Squibb,[113] self-injury in persons with autism did not improve significantly more in aripiprazole-treated children than in children in the placebo group. The improvement found in self-injury failed to reach the medical research standard of a $p$ value of 0.05 or lower, meaning that the study failed to show that the chance that the result was due to random factors, rather than to the drug, was 5 percent or less. In the flexible-dose study, there was only a 10 percent or less chance ($p < 0.10$) that the results showing improvement were random on the "injures self" component but not on the "hurts self" or "physical violence to self" components, and the fixed-dose study did not even have that significance, meaning that, except for the "injures self" component in the

111 James T. McCracken, James McGough, Bhavik Shah, Pegeen Cronin, Daniel Hong, Michael G. Aman, L. Eugene Arnold, et al., "Research Units on Pediatric Psychopharmacology Autism Network. Risperidone in Children with Autism and Serious Behavioral Problems," *New England Journal of Medicine* 347, no. 5 (2002): 314-21, doi:10.1056/NEJMoa013171; Sarah Shea, Atilla Turgay, Alan Carroll, Miklos Schulz, Herbert Orlik, Isabel Smith, and Fiona Dunbar, "Risperidone in the Treatment of Disruptive Behavioral Symptoms in Children with Autistic and Other Pervasive Developmental Disorders," *Pediatrics* 114, no. 5 (November 2004): e634-41, doi:10.1542/peds.2003-0264-F.

112 Katie Thomas, "J.&J. to Pay $2.2 Billion in Risperdal Settlement," *New York Times*, November 4, 2013, http://www.nytimes.com/2013/11/05/business/johnson-johnson-to-settle-risperdal-improper-marketing-case.html?_r=0).

113 Randall Owen, Linmarie Sikich, Ronald N. Marcus, Patricia Corey-Lisle, George Manos, Robert D. McQuade, William H. Carson, et al., "Aripiprazole in the Treatment of Irritability in Children and Adolescents with Autistic Disorder," *Pediatrics* 124, no. 6 (December 2009): 1540, doi:10.1542/peds.2008-3782; Ronald N. Marcus, Randall Owen, Lisa Kamen, George Manos, Robert D. McQuade, William H. Carson, and Michael G. Aman, "A Placebo-Controlled, Fixed-Dose Study of Aripiprazole in Children and Adolescents With Irritability Associated with Autistic Disorder," *Journal of the American Academy of Child & Adolescent Psychiatry* 48, no. 11 (November 2009): 1110, doi:10.1097/CHI.0b013e3181b76658.

flexible-dose study, the studies failed to show that the chance the results were due to random factors, rather than to the drug, was 10 or less.[114] The exact $p$ values of the individual items of the ABC-I were not published.

Furthermore, in the studies, there were side effects that were not controlled for and could have accounted for a decrease in the problem behaviors,[115] such as sedation (21 percent aripiprazole/4 percent placebo), somnolence (drowsiness) (10 percent aripiprazole/4 percent placebo), fatigue (17 percent aripiprazole/2 percent placebo), or lethargy (5 percent aripiprazole/0 percent placebo).[116] When people are sedated, they are not going to yell, scream, throw tantrums, or hurt others. They will also not learn and may not even be able to go down the stairs in the case of a fire. Sometimes behaviors cycle: they can get more intensive and more frequent and then less so; sometimes a cycle of improvement will just happen to coincide with medication, and we are not sure if the improvement is due to medication or it was a coincidence. This becomes a particular problem when the behavior intensity and frequency reoccur, but it is not clear if it is due to medication tolerance or if the initial improvement in behavior that started with the medication was simply a coincidence. Nevertheless, the FDA approval was still granted and included irritability with self-injurious behavior in autism.

In one placebo-controlled study with eighty-six subjects with intellectual disability, randomized to placebo, Haldol (haloperidol) or Risperdal (risperidone), the conclusion was as follows: "Antipsychotic drugs should no longer

114 Michael G. Aman, William Kasper, George Manos, Suja Mathew, Ronald Marcus, Randall Owen, and Raymond Mankoski, "Line-Item Analysis of the Aberrant Behavior Checklist: Results from Two Studies of Aripiprazole in the Treatment of Irritability Associated with Autistic Disorder," *Journal of Child and Adolescent Psychopharmacology* 20, no. 5 (October, 2010): 418, doi:10.1089/cap.2009.0120.

115 Owen, Skich, Marcus, Corey-Lisle, Manos, McQuade, Carson, et al., "Aripiprazole in the Treatment of Irritability in Children and Adolescents with Autistic Disorder," 1533-40; Marcus, Owen, Kamen, Manos, McQuade, Carson, and Aman, "A Placebo-Controlled, Fixed-Dose Study of Aripiprazole in Children and Adolescents with Irritability Associated with Autistic Disorder," 1110-19.

116 "Abilify US Full Prescribing Information," Otsuka Pharmaceutical, Tokyo, Japan, revised January 2016, 6.1, Table 20, http://www.otsuka-us.com/Documents/Abilify.PI.pdf.

be regarded as an acceptable routine treatment for aggressive challenging behaviour in people with intellectual disability."[117]

Furthermore, antipsychotics block the transmission of dopamine in the brain,[118] although Abilify (aripiprazole), Rexulti (brexpiprazole), and Vraylar (cariprazine), known as partial dopamine agonists, can enhance as well as block transmission, but at higher doses, they function as dopamine blockers.[119] Dopamine is needed for motivation,[120] which is what makes positive reinforcement effective and necessary to change behavior and even make rewards an incentive to do well at school. By blocking dopamine, these medications can potentially interfere with applied behavior analysis and learning in general by taking away motivation for positive behaviors.

In addition, some antipsychotics also block serotonin transmission.[121] As discussed earlier, brains in boys with autism have been found to have areas of low serotonin, although one area of the brain showed elevated serotonin.[122] Potentially, by administering these antipsychotics and thereby interfering with serotonin transmission, there may be potentially increased impairment in areas of the brain with reduced serotonin.

Regarding antidepressants (sometimes also used for anxiety), the following article shows that the class of antidepressants-selective serotonin reuptake inhibitors (SSRIs)-had limited utility in autism:

---

117 Peter Tyrer, Patricia C. Oliver-Africano, Zed Ahmed, Nick Bouras, Sherva Cooray, Shoumitro Deb, Declan Murphy, et al., "Risperidone, Haloperidol, and Placebo in the Treatment of Aggressive Challenging Behaviour in Patients with Intellectual Disability: A Randomised Controlled Trial," Lancet 371, no. 9606 (January 5, 2008): 57, doi:10.1016/S0140-6736(08)60072-0.

118 Benjamin James Sadock, Virginia Alcott Sadock, and Pedro Ruiz, Kaplan and Sadock's Synopsis of Psychiatry, 11th ed. (Philadelphia, PA: Wolters Kluwer, 2015), 1023.

119 Christoph U. Correll, "Advances in Dopamine Partial Agonism," Psych Evidence-Based Consults 1, no. 1 (November 2015): 11.

120 John D. Salamone and Merce Correa, "Review, the Mysterious Motivational Functions of Mesolimbic Dopamine," Neuron 76, no. 3 (November 8, 2012): 470, doi:10.1016/j.neuron.2012.10.021.

121 Sadock, Alcott Sadock, and Ruiz, *Kaplan and Sadock's Synopsis of Psychiatry*, 11th ed., 1023.

122 Chugani, Muzik, Rothermel, Behen, Chakraborty, Mangner, da Silva, et al., "Altered Serotonin Synthesis in the Dentatothalamocortical Pathway in Autistic Boys," 666-69.

> Nine RCTs [randomized controlled trials] with a total of 320 participants were included. Four SSRIs were evaluated: fluoxetine [Prozac] (three studies), fluvoxamine [Luvox] (two studies), fenfluramine [Pondimin] (two studies) and citalopram [Celexa] (two studies). Five studies included only children and four studies included only adults.

The author's conclusions were as follows:

> There is no evidence of effect of SSRIs in children and emerging evidence of harm. There is limited evidence of the effectiveness of SSRIs in adults from small studies in which risk of bias is unclear.[123]

Revia (naltrexone) is also used off label for self-injurious behavior. The FDA-approved indication for naltrexone is for alcohol and opiate dependence. Naltrexone blocks opiate receptors[124] and possibly can cause the individual to feel pain when exhibiting self-injury, which is why it is used. Therefore, the mechanism of use is as a possible aversive. However, the largest placebo-controlled trial showed no improvement in self-injury,[125] and naltrexone can cause liver toxicity, although this is not likely in recommended doses.[126] Regarding other side effects, "[s]tudies have reported that 25% to 65% of patients with schizophrenia have bone loss after taking antipsychotic drugs. Bone fractures in people with schizophrenia taking antipsychotics also occur more frequently than in the nonpsychiatric population."[127]

---

123 Katrina Williams, Amanda Brignell, Melinda Randall, Natalie Silove, and Philip Hazell, "Selective Serotonin Reuptake Inhibitors (SSRIs) for Autism Spectrum Disorders (ASD)," *Cochrane Database of Systematic Reviews* 8, no. CD004677 (August 20, 2013), doi:10.1002/14651858.CD004677.pub3.

124 Sadock, Alcott Sadock, and Ruiz, *Kaplan and Sadock's Synopsis of Psychiatry*, 11th ed., 1163.

125 Aynur Gormez, Fareez Rana, Susan Varghese, "Pharmacological Interventions for Self-Injurious Behaviour in Adults with Intellectual Disabilities," *Cochrane Database of Systematic Reviews* 4, CD009084 (April 30, 2013), doi:10.1002/14651858.CD009084.pub2.

126 Sadock, Alcott Sadock, and Ruiz, *Kaplan and Sadock's Synopsis of Psychiatry*, 11th ed., 638.

127 Ravinderjit Singh and Gagan Deep Mall, "Hyperprolactinemia in Antipsychotic Use," *Psychiatric Annals* 42, no. 10 (October 2012): 390, doi:10.3928/00485713-20121003-08.

Sometimes side effects of medications are not discovered for many years after a given medication has been FDA approved. For example, when Zyprexa (olanzapine) and other atypical antipsychotics were initially approved, physicians did not routinely check lipids or glucose to monitor for diabetes, so it took many years before it was known these could cause diabetes and elevated cholesterol and triglycerides. Later research showed that "[a]ntipsychotic medications appear to triple the risk of type 2 diabetes in children and teens, with most of the risk occurring in the first year of administration."[128] According to an analysis of thirteen studies involving 185,105 individuals, just being on the antipsychotic for at least three months appeared to double to triple the risk of diabetes in youth, and having an autism spectrum disorder increased this risk.[129] It was more than fifteen years after these medications were on the market that it was discovered that "[t]he atypical antipsychotics quetiapine (Seroquel), risperidone (Risperdal), and olanzapine (Zyprexa)...are associated with increased risk of acute kidney injury in older adults, suggests a study published last week in the *Annals of Internal Medicine*."[130] In addition, a meta-analysis found that antipsychotic use is associated with almost double the risk

128 Michele G. Sullivan, "Schizophrenia and Psychosis, Antipsychotics Triple Risk of Type 2 Diabetes in Young People," *Clinical Psychiatry News*, August 21, 2013, http://www.clinicalpsychiatrynews.com/single-view/antipsychotics-triple-risk-of-type-2-diabetes-in-young-people/e91193377ff41033df3357a3aa095aa3.html; William V. Bobo, William O. Cooper, C. Michael Stein, Mark Olfson, David Graham, James Daugherty, Catherine Fuchs, et al., "Antipsychotics and the Risk of Type 2 Diabetes Mellitus in Children and Youth," *JAMA Psychiatry* 70, no. 10 (October 2013): 1067-75. doi:10.1001/jamapsychiatry.2013.2053.

129 Britta Galling, Alexandra Roldán, René E. Nielsen, Jimmi Nielsen, Tobias Gerhard, Maren Carbon, Brendon Stubbs, et al., "Type 2 Diabetes Mellitus in Youth Exposed to Antipsychotics, a Systemic Review and Meta-analysis," *JAMA Psychiatry* 73, no. 3 (March 2016): 247-259, doi:10.1001/jamapsychiatry.2015.2923.

130 Jolynn Tumolo, "Atypical Antipsychotics Associated with Acute Kidney Injury in Older Adults," *Psych Congress Network*, August 26, 2014, http://www.psychcongress.com/article/atypical-antipsychotics-associated-acute-kidney-injury-older-adults-18655; Y. Joseph Hwang, Stephanie N. Dixon, Jeffrey P. Reiss, Ron Wald, Chirag R. Parikh, Sonja Gandhi, Salimah Z. Shariff, et al., "Atypical Antipsychotic Drugs and the Risk for Acute Kidney Injury and Other Adverse Outcomes in Older Adults: A Population-Based Cohort Study," *Annals of Internal Medicine* 161, no. 4 (August 19, 2014): 242-48, doi:10.7326/M13-2796.

of a heart attack,[131] and the risk is "especially among patients with schizophrenia or with drug use during the first 30 days."[132]

Similarly, Mellaril (thioridazine) was prescribed for many years without associated heart monitoring with electrocardiograms. My brother Matthew was one such recipient. We know now Mellaril can cause a fatal heart rhythm.[133] On June 18, 2013, the FDA published information about two deaths it was investigating after these individuals received Zyprexa injections.[134] Zyprexa has been on the market since 1996.[135] Despite its being on market for more than fifteen years, in February 2013, the FDA issued this warning on Zoloft (sertraline), an antidepressant: "ADVERSE REACTIONS, Other Events Observed During the Post marketing Evaluation, diabetes mellitus."[136] A recent study showed shrinking of the brain in individuals taking antipsychotics.[137] Furthermore, in 2015, evidence suggested that long-term exposure to antipsychotics in children "may decrease social functioning among young

---

131 Zheng-he Yu, Hai-yin Jiang, Li Shao, Yuan-yue Zhou, Hai-yan Shi, and Bing Ruan, "Use of Antipsychotics and Risk of Myocardial Infarction: A Systematic Review and Meta-analysis," *British Journal of Clinical Pharmacology* 82, no. 3 (September 2016): 624-32, doi:10.1111/bcp.12985.

132 Yu, Jiang, Shao, Zhou, Shi, and Ruan, "Use of Antipsychotics and Risk of Myocardial Infarction: A Systematic Review and Meta-analysis," 630.

133 Sadock, Alcott Sadock, and Ruiz, *Kaplan and Sadock's Synopsis of Psychiatry*, 11th ed., 914.

134 "Zyprexa Relprevv (Olanzapine Pamoate): Drug Safety Communication-FDA Investigating Two Deaths Following Injection," Safety Alerts for Human Medical Products, US Food and Drug Administration, updated March 23, 2015,
http://www.fda.gov/Safety/MedWatch/SafetyInformation/SafetyAlertsforHumanMedicalProducts/ucm357601.htm.

135 "Olanzapine: Drug Safety Communication - FDA Warns About Rare But Serious Skin Reactions," US Food and Drug Administration, May 10, 2016, http://www.fda.gov/Safety/MedWatch/SafetyInformation/SafetyAlertsforHumanMedicalProducts/ucm500123.htm.

136 "Zoloft (sertraline hydrochloride) Tablets and Oral Concentrate" Drug Safety Labeling Changes, US Food and Drug Administration, February 2013, http://www.fda.gov/safety/medwatch/safetyinformation/safety-relateddruglabelingchanges/ucm116470.htm.

137 Antonio Vita, Luca De Peri, Giacomo Deste, Stefano Barlati, and Emilio Sacchetti, "The Effect of Antipsychotic Treatment on Cortical Gray Matter Changes in Schizophrenia: Does the Class Matter? A Meta-analysis and Meta-regression of Longitudinal Magnetic Resonance Imaging Studies," *Biological Psychiatry* 78, no. 6 (September 15, 2015): 403-12, doi:10.1016/j.biopsych.2015.02.008.

adults,"[138] and social functioning is one of the core impairments in autism. In addition, another study showed that by giving young rats either Risperdal or Zyprexa, they exhibited hyperactivity later in life.[139] In the same study, male adult rats also exhibited behaviors indicative of anxiety after receiving either Risperdal, Zyprexa, or Abilify in their early lives. Both hyperactivity and anxiety are already very prevalent in individuals with autism. Other research conducted in 2015 showed that antidepressants, medications that have been on the market for decades, are linked to an increased risk of bleeding in the brain,[140] and individuals with autism already have neurological impairments. In addition, another study has shown that some antidepressants are associated with an increased incidence of dementia,[141] although depression itself may be a risk factor for dementia.[142]

Benzodiazepines have been on the market since 1960.[143] In September 2014, published research found that "[t]aking benzodiazepines for three months or longer is significantly associated with an increased risk-between

138 "Alternative Therapies Should Be Considered Before Antipsychotics for Children, Report Says," *American Psychiatric Association Psychiatric News Alert*, January 7, 2016, http://alert.psychnews.org/2016/01/alternative-therapies-should-be.html; William B. Daviss, Erin Barnett, Katrin Neubacher, and Robert E. Drake, "Use of Antipsychotic Medications for Nonpsychotic Children: Risks and Implications for Mental Health Services," *Psychiatric Services* 67, no. 3 (March 1, 2016): 339-41, doi:10.1176/appi.ps.201500272.

139 Michael De Santis, Jiamei Lian, Xu-Feng Huang, and Chao Deng, "Early Antipsychotic Treatment in Childhood/Adolescent Period Has Long-Term Effects on Depressive-like, Anxiety-like and Locomotor Behaviours in Adult Rats," *Journal of Psychopharmacology* 30, no. 2 (February 2016): 204-14, doi:10.1177/0269881115616383.

140 Lorraine L. Janeczko, "Antidepressant Use Linked to Increased Risk of Microbleeds," *Psych Congress Network,* January 6, 2016, http://www.psychcongress.com/article/antidepressant-use-linked-increased-risk-microbleeds-25851; Saloua Akoudad, Nikkie Aarts, Raymond Noordam, M. Arfan Ikram, Henning Tiemeier, Albert Hofman, Bruno H. Stricker, et al., "Antidepressant Use Is Associated with an Increased Risk of Developing Microbleeds," *Stroke* 47, no. 1 (January 2016): 251-54, doi:10.1161/STROKEAHA.115.011574.

141 Cynthia Wei-Sheng Lee, Cheng-Li Lin, Fung-Chang Sung, Ji-An Liang, and Chia-Hung Kao, "Antidepressant Treatment and Risk of Dementia: A Population-Based, Retrospective Case-Control Study," *Journal of Clinical Psychiatry* 77, no. 1 (2016): 117-22, doi:10.4088/JCP.14m09580.

142 Ibid., 118.

143 Jeannette Y. Wick, "The History of Benzodiazepines," *Consult Pharmacist* 28, no. 9 (September 2013): 538, doi:10.4140/TCP.n.2013.538.

43 to 51%-of developing Alzheimer's disease."[144] However, a later study did not find that long-term benzodiazepine use was linked to dementia, making the risk uncertain, although in the study it was noted that "[f]ew participants had heavy benzodiazepine use."[145] Further research showed that "[o]f the 11 studies published on benzodiazepines and risk of dementia, there were 9 that found that the drugs have a harmful effect."[146] In addition, in 2014, another study that followed individuals over seven years found an increased mortality in subjects taking benzodiazepines and other antianxiety and hypnotic drugs.[147]

Weight gain is also one of the side effects of most psychotropic medications. "Several evidence-based studies have shown that obese teens have a higher incidence of mental health problems such as depression, anxiety and poor self-esteem than nonobese teens."[148]

---

144 Jolynn Tumolo, "Benzodiazepines Associated with Increased Alzheimer's Risk," *Psych Congress Network,* September 19, 2014, http://www.psychcongress.com/article/benzodiazepines-associated-increased-alzheimer%E2%80%99s-risk-18903; Sophie Billioti de Gage, Yola Moride, Thierry Ducruet, Tobias Kurth, Helene Verdoux, Marie Tournier, Antoine Pariente, et al., "Benzodiazepine Use and Risk of Alzheimer's Disease: Case-Control Study," *British Medical Journal* 349, g5205 (September 9, 2014), doi:10.1136/bmj.g5205; Science Daily, "Long Term Use of Pills for Anxiety and Sleep Problems May Be Linked to Alzheimer's," September 9, 2014, http://www.sciencedaily.com/releases/2014/09/140909192042.htm; Kristine Yaffe and Malaz Boustani, "Benzodiazepines and Risk of Alzheimer's Disease," *British Medical Journal* 349, g5312 (September 9, 2014), doi:10.1136/bmj.g5312.

145 Shelly L. Gray, Sascha Dublin, Onchee Yu, Rod Walker, Melissa Anderson, Rebecca A. Hubbard, Paul K. Crane, et al., "Benzodiazepine Use and Risk of Incident Dementia or Cognitive Decline: Prospective Population Based Study," *British Medical Journal* 352, i90 (February 2, 2015): 8, doi:10.1136/bmj.i90.

146 Lauren LeBano, "Update on the Link between Benzodiazepines and Dementia," *Psych Congress Network,* February 22, 2016, http://www.psychcongress.com/article/update-link-between-benzodiazepines-and-dementia-26540; Antoine Pariente, Sophie Billioti de Gage, Nicholas Moore, and Bernard Bégaud, "The Benzodiazepine-Dementia Disorders Link: Current State of Knowledge," *CNS Drugs* 30, no. 1 (January 2016): 1-7, doi:10.1007/s40263-015-0305-4.

147 Scott Weich, Hannah Louise Pearce, Peter Croft, Swaran Singh, Ilana Crome, James Bashford, and Martin Frisher," Effect of Anxiolytic and Hypnotic Drug Prescriptions on Mortality Hazards: Retrospective Cohort Study," *British Medical Journal* 348, g1996 (March 19, 2014): 1-12, doi:10.1136/bmj.g1996.

148 Deina Nemiary, Ruth Shim, Gail Mattox, and Kisha Holden, "The Relationship Between Obesity and Depression among Adolescents," *Psychiatric Annals* 42, no. 8 (2012): 305, doi:10.3928/00485713-20120806-09.

Additionally, "[i]n a patient with ASD, depression frequently presents as an increase of existent ASD symptoms. What is important to note is not the symptoms themselves, but their heightened intensity."[149] This heightened intensity may result in more medication, which is a vicious cycle. In addition, obesity in general is associated with a severe course of depression and structural brain alterations.[150]

Furthermore, medications can cause discomforting and even painful side effects-such as constipation-that may manifest in increased agitation and dangerous behavior. Medications can be particularly dangerous for individuals with limited communication skills, as they may not be able to disclose symptoms of their side effects, which may allow adverse effects to become life threatening before they are detected. When they are detected, simply stopping the drug does not remove the drug from the body immediately, and stopping a drug immediately can create withdrawal side effects.

Individuals with developmental disabilities may be more susceptible to side effects as well. For example, there is a "greater sensitivity to somnolence [drowsiness] and sedation in children and adolescents (e.g., in 22-30% of subjects taking risperidone [Risperdal] for pediatric mania and in 49% of risperidone monotherapy-treated children in FDA registration trials for autism, in contrast to an incidence of <7% in adults with schizophrenia or of 5% in adults with bipolar mania taking risperidone)."[151] In fact, in one of the FDA registration trials for autism, the prevalence of somnolence (drowsiness) was 72.5 percent versus 7.7 percent for placebo.[152]

---

149 Zeynep Ozinci, Tara Kahn, and Laura N. Antar, "Depression in Patients with Autism Spectrum Disorder," *Psychiatric Annals* 42, no. 8 (August 2012): 294, doi:10.3928/00485713-20120806-06.

150 Nils Opel, Ronny Redlich, Dominik Grotegerd, Katharina Dohm, Walter Heindel, Harald Kugel, Volker Arolt, et al., "Obesity and Major Depression: Body-Mass Index (BMI) Is Associated with a Severe Course of Disease and Specific Neurostructural Alterations," *Psychoneuroendocrinology* 51 (January 2015): 219-26, doi:10.1016/j.psyneuen.2014.10.001.

151 Joseph F. Goldberg and Carrie L. Ernst, *Managing the Side Effects of Psychotropic Medications* (Washington, DC: American Psychiatric Publishing, 2012), 72.

152 Shea, Turgay, Carroll, Schulz, Orlik, Smith, and Dunbar, "Risperidone in the Treatment of Disruptive Behavioral Symptoms in Children with Autistic and Other Pervasive Developmental Disorders," e639.

In addition, the measured improvements in the behaviors in only one of the FDA registration trials controlled for sedation,[153] although in both studies, the majority of the cases of sedation did resolve with time.[154]

Also, these side effects can considerably impair functioning and damage an individual's quality of life. Consider a child with autism who develops diabetes from psychotropic medications. This child already has communication and social impairments. Now, if this child goes to a birthday party and cannot understand why he or she cannot eat the pizza and cake or can only have a limited amount, the child will not have a positive experience. Individuals with limited communication skills may never understand why their fingers are getting pricked multiple times a day to check their blood sugars or why they are receiving insulin injections.

In addition, diabetes can contribute to cerebral atrophy (destruction of brain cells):

> In addition to increasing patients' risk for cardiovascular disease, stroke, and cancer, obesity and metabolic disturbance contribute to age-related cognitive decline and dementia. In particular, insulin resistance and hyperinsulinemia promote neurocognitive dysfunction and neurodegenerative changes during the extended, preclinical phase of Alzheimer's disease…BMI [body mass index]…and fasting insulin are positively correlated with atrophy in frontal, temporal, and subcortical brain regions.[155]

---

153 Ibid., e639.

154 McCracken, McGough, Shah, Cronin, Hong, Aman, Arnold, et al., "Research Units on Pediatric Psychopharmacology Autism Network. Risperidone in Children with Autism and Serious Behavioral Problems, 318-19; Shea, Turgay, Carroll, Schulz, Orlik, Smith, and Dunbar, "Risperidone in the Treatment of Disruptive Behavioral Symptoms in Children with Autistic and Other Pervasive Developmental Disorders," e638-e639.

155 Robert Krikorian, "Metabolic Disturbance and Dementia: A Modifiable Link," *Current Psychiatry* 12, no. 3 (March 2013): 17-18, http://www.mdedge.com/currentpsychiatry/article/65040/alzheimers-cognition/metabolic-disturbance-and-dementia-modifiable.

Even without dementia, type 2 diabetes is linked to low performance on cognitive tests.[156] Cognition is already impaired in individuals with developmental disabilities. Furthermore, one article noted as follows:

> In the presence of diabetes, the prevalence of depression increases to 15% to 30% depending on depression definition, population sample, and study type…In the case of depression, changes in blood sugar levels have been linked directly to moods such as anger, anxiety, sadness, frustration, and even general well-being-common themes in depressed patients. Clinically, more than 70% of patients with diabetes have depressive episodes that last longer than 2 years.[157]

In addition, one study stated, "[t]he incident AD [affective disorder] risk is increased by 2.6-fold in T2DM [type 2 diabetes mellitus]."[158] Furthermore, "[d]iabetes with fluctuating sugars has long been known to cause insomnia."[159] The resulting insomnia in itself or the resulting agitation may lead to further psychotropic medication use.

---

156 Alessia D'Anna, "Type 2 Diabetes Linked to Decreased Cognitive Function," *Psych Congress Network*, March 9, 2015, http://www.psychcongress.com/article/type-2-diabetes-linked-decreased-cognitive-function-21501; Corita Vincent and Peter A. Hall, "Executive Function in Adults with Type 2 Diabetes: A Meta-Analytic Review," *Psychosomatic Medicine* 77, no. 6 (July/August 2015): 631-42, doi:10.1097/PSY.0000000000000103; "Type 2 Diabetes Linked to Worse Performance on Cognitive Testing," EurekAlert!, February 13, 2015, http://www.eurekalert.org/pub_releases/2015-02/uow-t2d021315.php.

157 Shane M. Coleman and Wayne J. Katon, "Treatment Implications for Comorbid Diabetes Mellitus and Depression," *Psychiatric Times*, January 18, 2013, http://www.psychiatrictimes.com/treatment-implications-comorbid-diabetes-mellitus-and-depression.

158 Mark L Wahlqvist, Meei-Shyuan Lee, Shao-Yuan Chuang, Chih-Cheng Hsu, Hsin-Ni Tsai, Shu-Han Yu, and Hsing-Yi Chang, "Increased risk of affective disorders in type 2 diabetes is minimized by sulfonylurea and metformin combination: a population-based cohort study," *BMC Medicine* 10, no. 150 (2012): 1, doi:10.1186/1741-7015-10-150.

159 Richard C. Holbert, Khurshid A. Khurshid, Robert Averbuch, and Imran S. Khawaja, "Sleep and Schizophrenia," *Psychiatric Annals* 46, no. 3 (March 2016): 195, doi:10.3928/00485713-20160205-01.

When individuals have increased appetites and become overweight, obese and/or prediabetic or diabetic from their medications, residences put them on diets. These individuals often become agitated and have dangerous behaviors over food, prompting increases in medications, again creating a vicious cycle. Some staff members will simply allow the patients to eat foods that are contraindicated for their medical conditions, because they do not want to and cannot safely deal with their charges' dangerous behaviors. Individuals who are more verbal, and in some residences even individuals who are nonverbal, are simply allowed to eat whatever they choose.

New York State Medicaid does not pay for visits with a registered dietician for being overweight (BMI 25-29.9) except for a one-time initial evaluation. Patients must be clearly obese (BMI 30 or higher) to qualify for ongoing monitoring. Being overweight increases the risk for medical problems such as heart disease, diabetes, or cancer. For example, being overweight without obesity will increase the risk of dying of breast cancer in women by 34 percent and the risk of dying of uterine cancer by 50 percent, and the risk of some other cancers is also increased in men and women.[160] The ten-year risk of developing diabetes in individuals who are overweight is 4.6 and 3.5 times normal weight people in females and males respectively.[161] In addition, once weight is gained, it is difficult to lose. In October 2015, New York State Medicaid changed the funding source for individuals who live in residential settings to receive nutrition services through the agency in which they reside rather than directly through Medicaid. However, as of March 2016, some agencies have not been paying for these consults, resulting in some individuals not being served. Even when some agencies hired dieticians, residents lost individual nutritional counseling. Refusing to fund

160 Eugenia E. Calle, Carmen Rodriguez, Kimberly Walker-Thurmond, and Michael J. Thun, "Overweight, Obesity, and Mortality from Cancer in a Prospectively Studied Cohort of U.S. Adults," *New England Journal of Medicine* 348, no. 17 (April 24, 2003): 1628-31, doi:10.1056/NEJMoa021423.

161 Alison E. Field, Eugenie H. Coakley, Aviva Must, Jennifer L. Spadano, Nan Laird, William H. Dietz, Eric Rimm, et al., "Impact of Overweight on the Risk of Developing Common Chronic Diseases During a 10-Year Period," *JAMA Archives of Internal Medicine* 161, no. 13 (July 9, 2001): 1584, doi:10.1001/archinte.161.13.1581.

dietician consults will only result in people becoming sicker, more medication being prescribed, and increased costs.

Some individuals with developmental disabilities do not understand and will not comply with the necessary medical monitoring for such things as blood pressure, or for blood tests to check for cholesterol levels, diabetes, liver problems, or other potential adverse side effects. Blood tests are painful and, unlike the skin-shock treatment, the pain lasts longer than two seconds. Clozaril (clozapine), an antipsychotic sometimes used to manage behavior these individuals, requires weekly blood tests for six months, then blood work once every two weeks for another six months, and then monthly testing. The Abnormal Involuntary Movement Scale, which is recommended every three to six months for individuals on antipsychotics to screen for tardive dyskinesia, cannot be completed in some individuals who do not have the cognitive skills to follow directions. Sometimes they need to be physically restrained for blood tests, which carries inherent risks.

Once, Talia was prescribed Zofran (ondansetron) for intractable vomiting due to a virus. I did not administer the medication at night, because a rare but deadly side effect is neuroleptic malignant syndrome (a potential side effect of all antipsychotics), for which Matthew previously experienced symptoms. I knew if she had symptoms during the night, she could not communicate them to me. I cannot help thinking of the risk in prescribing antipsychotics to individuals with communication impairments who can die in the middle of the night while parents are fast asleep.

Psychiatric medications can also have adverse psychiatric effects that a disabled individual may not be able to communicate. For example,

> Psychosis is a known risk of psychostimulants, usually in a dose-related fashion...[r]are case reports of new hallucinations...have been described in association with the use of some SSRIs (notably, fluvoxamine) or SNRIs (venlafaxine)...Cases also exist of new-onset psychosis associated with bupropion [Wellbutrin], typically within days to weeks of treatment initiation, and observed even at relatively low dosages (100-150 mg/day).[162]

162 Goldberg and Ernst, *Managing the Side Effects of Psychotropic Medications*, 102.

"Epilepsy is increasingly recognized as the most common medical disorder accompanying ASD."[163] Furthermore, up to 60 percent of children with autism without clinical seizures have epileptiform discharges on electroencephalograms (EEGs) according to published research.[164] Most psychotropic medications will increase their risk for seizures when many individuals already have an increased risk. As stated earlier, this occurred with Stuart.

One amino acid (a protein building block), homocysteine, may be involved in autism, and some psychiatric medication may increase homocysteine levels further. "Areas of ongoing investigation in which the toxic and genetic stress of HCY [homocysteine] are noted to play a role in pathogenesis include...autism."[165] Mood stabilizers such as Lamictal (lamotrigine), Tegretol (carbamazepine), and lithium can further increase homocysteine levels, and even moderately elevated levels are associated with vascular disease and increased mortality.[166] Furthermore, some toxic levels of homocysteine are associated with thyroid disease, renal failure, chronic inflammatory conditions, malignancies, gastrointestinal disorders, and poor vitamin absorption.[167]

For medication in individuals over the age of forty, one study of more than 332 subjects concluded the following:

---

163 Fry Williams and Lee Williams, *Effective Programs for Treating Autism Spectrum Disorder: Applied Behavior Analysis Models*, 14; Roberto Canitano, "Epilepsy in Autism Spectrum Disorders," *European Child & Adolescent Psychiatry* 16, no. 1 (2007): 61-66, doi:10.1007/s00787-006-0563-2; Fred R. Volkmar, Catherine Lord, Anthony Bailey, Robert T. Schultz, and Ami Klin, "Autism and Pervasive Developmental Disorders," *Journal of Child Psychology and Psychiatry* 45, no. 1 (January 2004): 135-70, doi:10.1046/j.0021-9630.2003.00317.x.

164 Sarah J. Spence and Mark T. Schneider, "The Role of Epilepsy and Epileptiform EEGs in Autism Spectrum Disorders," *Pediatric Research* 65, no. 6 (June 2009): 599, doi:10.1203/01.pdr.0000352115.41382.65.

165 Angela Pana, "Homocysteine and Neuropsychiatric Disease," *Psychiatric Annals* 45, no. 9 (September 2015): 467, doi:10.3928/00485713-20150901-05.

166 Ibid, 465.

167 Ibid.

> Employing a study design that closely mimicked clinical practice, we found a lack of effectiveness and a high incidence of side effects with 4 commonly prescribed atypical antipsychotics across diagnostic groups in patients over age 40, with relatively few differences among the drugs. Caution in the use of these drugs is warranted in middle-aged and older patients…Caution is needed in long-term use of commonly prescribed atypical antipsychotics (aripiprazole, olanzapine, quetiapine, and risperidone) in middle-aged and older patients with psychotic disorders. When these medications are used, they should be given in low dosages, for short durations, and their side effects should be monitored closely. Shared decision making with patients and their caregivers is recommended, including discussions of risks and benefits of atypical antipsychotics and those of available treatment alternatives.[168]

Contrary to the findings of this article, in the developmental disabilities population, these medications are often given at high dosages, for the long term, and it is difficult to monitor side effects due to communication impairments.

As far as shared decision making goes, it is difficult for parents to refuse when the day program tells them their child will be discharged or the residence tells them they have to take their child home if they do not consent. Furthermore, if parents still refuse, hospitals can obtain involuntary court orders, and agencies can have parents' legal guardianship removed by stating that the parent is not acting in the interest of the individual who poses a danger, as occurred to my cousins, which is discussed elsewhere. However, there are never court orders for employing a board-certified behavior analyst or conducting a functional experimental analysis to find out why the behaviors are occurring or highly structured intensive behavior plans to reduce them. There is never a court order for a day program and residence to make sure they are working together for a consistent behavior plan.

168 Hua Jin, Pei-an Betty Shih, Shahrokh Golshan, Sunder Mudaliar, Robert Henry, Danielle K. Glorioso, Stephan Arndt, et al., "Comparison of Longer-Term Safety and Effectiveness of Four Atypical Antipsychotics in Patients Over Age 40: A Trial Using Equipoise-Stratified Randomization," *Journal of Clinical Psychiatry* 74, no. 1 (January 2013): 10-11, doi:10.4088/JCP.12m08001.

Sometimes at an annual meeting for an individual at a day program, no one from the residence even shows up. The consistency is further complicated as many individuals attend day programs at agencies different than their residences.

The government, as well as private insurance companies, sometimes prefers doctors to prescribe one medication over another. Insurance companies have formularies of preferred drugs. In 2007 and 2008, Medicaid in New York wanted physicians to prescribe Ambien (zolpidem) CR for insomnia, manufactured by Pfizer. However, if prescribers wanted to prescribe a much cheaper generic zolpidem, they had to call Medicaid to get prior authorization, including a reason why the individual could not take Ambien CR.

In New York State, the government has quite a say in what medications are prescribed. In 2011, the state created an extensive Medicaid-preferred drug list. In my opinion, this list violated many of the tenets of good care by removing a physician's choice.

For example, particularly in an overweight diabetic, if an antipsychotic medication is needed, one should first try a medication with a lower risk for weight gain before prescribing one with a higher risk, which would be more likely to exacerbate diabetes. Depending on other individual variables, Abilify (aripiprazole) may be a better choice than quetiapine (generic) or Seroquel XR (brand), which has a higher likelihood of weight gain and is more likely to exacerbate diabetes. However, in 2014 and most of 2015, New York State Medicaid required prior authorization for Abilify explaining why the individual could not take a preferred drug, while Seroquel XR has been on the preferred drug list.[169]

New York State Medicaid paid $1,164.84, $1281.32, and $1383.83 for a thirty-day supply of Seroquel XR 800 mg (two 400 mg tablets) per day, the FDA-approved maximum dose, in 2013, 2014, and 2015 respectively.[170] As

169 New York State Fee-For-Service Pharmacy Programs, revised October 30, 2015, 14, https://newyork.fhsc.com/downloads/providers/nyrx_pdp_pdl.pdf. A prior authorization for a nonpreferred drug must provide an explanation why the nonpreferred drug is necessary, and the explanation must be approved.

170 "New York State Department of Health List of Medicaid Reimbursable Drugs," 2013, 2014, and 2015, https://www.emedny.org/info/fullform.pdf.

one article stated, "According to the International Federation of Health Plans, Americans pay anywhere from two to six times more than the rest of the world for brand name prescription drugs."[171] Furthermore, the federal Affordable Care Act, also known as Obamacare, forbids negotiating prices with pharmaceutical companies.[172]

The FDA's maximum daily dose of Abilify is only 30 mg daily. The dose is lower than Seroquel because it is a more potent drug. Medicaid paid $942.57, $1093.39, and $1,256 in 2013, 2014, and 2015, respectively, for a thirty-day supply of 30 mg tablets.[173] In summary, the costs of Abilify have been lower than the costs for Seroquel XR for FDA-approved maximum dosages. Furthermore, while Seroquel XR has been on the preferred drug list, the much cheaper generic quetiapine as well as Seroquel XR are only "preferred" at doses of 100 mg daily or greater,[174] even though according to the FDA treatment guideline adults with schizophrenia and bipolar depression and all children are supposed to have 50 mg for the first day.[175] Furthermore, in individuals with autism, a maintenance dose may need to be 75 mg daily.[176]

In addition, New York State early intervention providers have also had their pay rates cut in recent years, earning less than they did in 1993.[177] The pay cuts were never restored. Meanwhile, the pharmaceutical companies have

---

171 Nadia Kounang, "Why Pharmaceuticals Are Cheaper Abroad," *CNN*, September 28, 2015, http://www.cnn.com/2015/09/28/health/us-pays-more-for-drugs/.

172 Ed Silverman, "Will the Affordable Care Act Give Drugmakers a Boost?" Forbes, December 26, 2013, http://www.forbes.com/sites/edsilverman/2013/12/26/will-the-affordable-care-act-give-drugmakers-a-boost/#85fbffd39b0c.

173 "New York State Department of Health List of Medicaid Reimbursable Drugs," 2013, 2014, and 2015, https://www.emedny.org/info/fullform.pdf. Prices per tablets are given, which can be multiplied to obtain the monthly cost.

174 New York State Fee-For-Service Pharmacy Programs, revised October 30, 2015, 14 https://newyork.fhsc.com/downloads/providers/nyrx_pdp_pdl.pdf.

175 Labeling - Food and Drug Administration for quetiapine, http://www.accessdata.fda.gov/drugsatfda_docs/label/2009/020639s045s046lbl.pdf.

176 David J. Posey, Kimberly A. Stigler, Craig A. Erickson, and Christopher J. McDougle, "Treatment of Autism with Antipsychotics," in *Clinical Manual for the Treatment of Autism*, ed. Eric Hollander and Evdokia Anagnostou (Washington, DC: American Psychiatric Publishing, 2007), 115.

177 Cathleen F. Crowley, "Tots Suffer as Therapy Cut," *Albany Times Union*, June 6, 2011, http://www.timesunion.com/local/article/Tots-suffer-as-therapy-cut-1410967.php.

had substantial raises in their sales of prescription medications. Furthermore, if early intervention were funded better, children would improve significantly and should require less medication when they are older.

In any case, with the preferred drug list, the government and private insurance companies with agreements with certain pharmaceutical companies get to decide what is prescribed, not doctors. Some patients even lose access to their medications for a few days or even weeks if prior authorization is needed, and there is a problem obtaining it. This is especially problematic for nonpreferred drugs from Medicare Part D and Medicaid farmed-out private insurers. I had a patient off medication for days because the insurance company denied ever receiving my prior authorization form, although it had been faxed together with another prior authorization form they claimed they received. I have wasted hours on the phone with automations and transfers to obtain prior authorizations, something that is not easy to do when I have up to seventeen patients to be seen on any given day. It is not that they do not want you to prescribe medication, but that they want you to prescribe medications from their preferred partners. While Medicaid will pay for a psychiatrist, they will not pay for any BCBA for an individual service, and the legislative pushes for health insurance to fund ABA often do not extend to Medicaid.

I remember when one of my patient's insurance companies would no longer pay for Zyprexa Zydis (orally disintegrating olanzapine), and the residence told me that to buy a thirty-day supply privately would cost $650 a month for that one tablet a day. Now add that to the doctor's visits and the blood tests to monitor that medication. Further add that to the cost of diabetes, heart disease, liver problems, and other side effects that often arise, sometimes years later.

Others have found the trials lacking in side-effect information: "Pharmacotherapy trials undertaken by industry sponsors are usually driven by the pursuit of product indications from the U.S. Food and Drug Administration (FDA), rather than by initiatives to counteract (and draw attention to) the adverse effects of a proprietary drug-in other words, there is a dearth of systemic research on the treatment of side effects."[178] Additionally,

178 Goldberg and Ernst, preface to *Managing the Side Effects of Psychotropic Medications*, xviii.

"[n]umerous factors likely contribute to the so-called gap between treatment efficacy (i.e., optimal results) and effectiveness (i.e., customary results), including...the tendency for adverse drug effects to be registered passively and incompletely in randomized controlled trials."[179]

*Managing the Side Effects of Psychotropic Medications*, further stated, "Despite the rigors of all phases of safety testing for a drug, pharmacovigilance data at times emerge in ways that could not be anticipated initially in the absence of more extensive exposure to larger segments of the population."[180] The price of medications, including the doctor's visits, possible hospitalizations, and other treatments for the side effects, get added every year for the rest of that individual's life.

This process produces something much more expensive than a free and appropriate education, something much more expensive than the speech and language therapies that teach to communicate rather than act out, and something much more expensive than the behavioral therapies that decrease dangerous behaviors. Furthermore, ABA therapies can be much more effective and have longer-lasting results than medication. Being on cocktails of medications, as in Stuart's case, prescribed for a different indication than approved by the FDA, does not necessarily do what it is meant to do to stop dangerous behaviors.

If a child is engaging in aggression or self-injury because of a lack of communication, imagine what might be accomplished by increasing speech and language therapy for example from twice to five days a week for thirty minutes. This increase of three days a week in New York City would cost $45 a session, or $585 a month. Receiving this speech therapy to improve communication and reduce dangerous behavior, as opposed to receiving Seroquel XR 800 mg daily for aggression, would save $909.53 a month on the basis of the 2016 price New York State Medicaid pays for Seroquel XR, $1,494.53 for a thirty day supply of 800 mg daily.[181] It would also provide a

179 Goldberg and Ernst, *Managing the Side Effects of Psychotropic Medications*, 26.

180 Ibid., 37.

181 "New York State Department of Health List of Medicaid Reimbursable Drugs," 2016, https://www.emedny.org/info/fullform.pdf.

job, teach the child a skill, and unlike medication, would not be likely have to be given for life. Unlike medication, the speech therapy would also not cause adverse effects that are costly to treat and diminish the individual's quality of life.

### *Forced Medications*

In 2011, a supervisor informed me that to obtain services, such as day habilitation, service coordination, or after-school programming from the OPWDD for individuals diagnosed with Attention-Deficit Hyperactivity Disorder (ADHD), the individuals might be required to take medication for the ADHD for six months and be retested to see if the IQ still indicated an intellectual disability and if the Vineland Adaptive Behavior Scales indicated significant impairment. This is as if a few points' increase in an IQ score would indicate that the individual will not need services after all. Also, a recent review of 185 studies involving 12,245 children or adolescents with a diagnosis of ADHD-treated with methylphenidate, a stimulant used to treat ADHD-concluded that quality of the evidence in the research studies was low and suggested clinicians prescribe methylphenidate "only for a maximum of three months."[182] The stimulants used in ADHD medications have potential serious side effects such as seizures, weight loss, increased heart rate and blood pressure, palpitations, rebound hyperactivity, and precipitation of tic or other movement disorders, irritability, and agitation.[183] Furthermore, "[c]hildren and adolescents who take stimulants for attention-deficit hyperactivity disorder (ADHD) are at twice the risk of experiencing a cardiovascular event than children with ADHD not prescribed psychostimulants, according to a prospective study of

---

182 Nick Zagorski, "New Cochrane Review Urges Caution in Prescribing Methylphenidate to Children," *American Psychiatric Association, Psychiatric News*, December 4, 2015, http://psychnews.psychiatryonline.org/doi/full/10.1176/appi.pn.2015.PP12a4; Ole Jakob Storebø, Erica Ramstad, Helle B. Krogh, Trine Danvad Nilausen, Maria Skoog, Mathilde Holmskov, and Susanne Rosendal, "Benefits and Harms of Methylphenidate for Children and Adolescents with Attention Deficit Hyperactivity Disorder (ADHD)," *Cochrane Library* 11, no. CD009885 (November 25, 2015), doi:10.1002/14651858.CD009885.pub2.

183 Sadock, Alcott Sadock, and Ruiz, *Kaplan and Sadock's Synopsis of Psychiatry*, 11th ed., 1035-36.

more than 700,000 children in Denmark, 8,300 of whom had ADHD."[184] Another study involving children and adolescents showed an increased risk of heart attacks in the first two months of treatment with the stimulant methylphenidate, the active ingredient in Ritalin, Concerta, Focalin, Metadate, and other brand name stimulants.[185] Furthermore, according to the study, methylphenidate also increased the risk of heart rhythm abnormalities during different time periods after exposure to methylphenidate. Even though the absolute risk for these adverse events "is likely to be low," which means that, even with the medication, it is still unlikely to experience these side effects, there is an elevated risk when taking the drug compared to not taking the drug. In addition, amphetamines, the active ingredient in Adderall, Evekeo, Dyanavel, and other brand name stimulants, have been shown to cause death to brain cells in rats.[186] Methylphenidate also can cause brain damage in rats.[187] Whether this also occurs in humans is unknown.

Strattera (atomoxetine), also used with ADHD, has a black-box warning (a warning designed to call attention to serious or life-threatening risks) on the label regarding possible suicidality,[188] and Strattera can cause severe liver

---

184 Jolynn Tumolo, "Stimulants for ADHD Tied to Increased Cardiovascular Risk in Children," *Psych Congress Network*, July 16, 2014, http://www.psychcongress.com/article/stimulants-adhd-tied-increased-cardiovascular-risk-children-18181?e=[email]; Soren Dalsgaard, Anette Primdal Kvist, James F. Leckman, Helena Skyt Nielsen, and Marianne Simonsen, "Cardiovascular Safety of Stimulants in Children with Attention-Deficit/Hyperactivity Disorder: A Nationwide Prospective Cohort Study," *Journal of Child and Adolescent Psychopharmacology* 24, no.6 (August 19, 2014): 302-10, doi:10.1089/cap.2014.0020; "Does Psychostimulant Use Increase Cardiovascular Risk in Children with ADHD?" *EurekAlert!*, June 26, 2014.

185 Ju-Young Shin, Elizabeth E. Roughead, Byung-Joo Park, and Nicole L. Pratt, "Cardiovascular Safety of Methylphenidate among Children and Young People with Attention-Deficit/Hyperactivity Disorder (ADHD): Nationwide Self Controlled Case Series Study," *British Medical Journal* 353, i2550 (May 2016), doi:10.1136/bmj.i2550.

186 G. Stumm, J. Schlegel, T. Schafer, C. Wurz, H. D. Mennel, J. C. Krieg, and H. Vedder, "Amphetamines Induce Apoptosis and Regulation of bcl-x Splice Variants in Neocortical Neurons," *The Official Journal of the Federation of American Societies for Experimental Biology* 13, no. 9 (June 1999): 1065-72.

187 Marcio R. Martins, Adalisa Reinke, Fabricia C. Petronilho, Karin M. Gomes, Felipe Dal-Pizzol, and Joao Quevedo, "Methylphenidate Treatment Induces Oxidative Stress in Young Rat Brain," *Brain Research* 1078, no. 1 (March 17, 2006): 189-97, doi:10. 1016/j/brainres 2006.01.004.

188 *Strattera Medication Guide* (Indianapolis, IN: Eli Lilly and Company, 2002, 2015), http://www.fda.gov/ohrms/dockets/dockets/06p0209/06P-0209-EC3-Attach-1.pdf.

injury.[189] Kapvay (clonidine) and Intuniv (guanfacine), other drugs approved for ADHD, can lower blood pressure and decrease heart rate.[190] Yet the government seems to prefer to fund the development and treatment of medical problems rather than give someone a more productive life.

On another note, extensive controlled studies have linked attention deficits and hyperactive symptoms to food coloring.[191] One literature analysis suggests trying an elimination diet to treat ADHD symptoms, which includes

189 Sadock, Alcott Sadock, and Ruiz, *Kaplan and Sadock's Synopsis of Psychiatry*, 11th ed., 1038.

190 Ibid., 931.

191 Donna McCann, Angelina Barrett, Alison Cooper, Debbie Crumpler, Lindy Dalen, Kate Grimshaw, Elizabeth Kitchin, et al., "Food Additives and Hyperactive Behaviour in 3-Year-Old and 8/9-Year-Old Children in the Community: A Randomised Double-Blinded, Placebo-Controlled Trial," *Lancet* 370, no. 9598 (November 3, 2007): 1560-67, doi:10.1016/S0140-6736(07)61306-3; David W. Schab and Nhi-Ha T. Trinh, "Do Artificial Food Colors Promote Hyperactivity in Children with Hyperactive Syndromes? A Meta-Analysis of Double-Blind Placebo-Controlled Trials," *Journal of Developmental & Behavioral Pediatrics* 25, no. 6 (December 2004): 423-34, doi:10.1097/00004703-200412000-00007; Laura J. Stevens, Thomas Kuczek, John R. Burgess, Elizabeth Hurt, and L. Eugene Arnold, "Dietary Sensitivities and ADHD Symptoms: Thirty-five Years of Research," *Clinical Pediatrics* 50, no. 4 (April 2011): 279-93, doi:10.1177/0009922810384728 ; Joel T. Nigg, Kara Lewis, Tracy Edinger, and Michael Falk, "Meta-Analysis of Attention-Deficit/Hyperactivity Disorder or Attention-Deficit/Hyperactivity Disorder Symptoms, Restriction Diet, and Synthetic Food Color Additives," *Journal of the American Academy of Child & Adolescent Psychiatry* 51, no. 1 (January 2012): 86-97, doi:10.1016/j.jaac.2011.10.015; L. Eugene Arnold, Nicholas Lofthouse, and Elizabeth Hurt, "Artificial Food Colors and Attention-Deficit/Hyperactivity Symptoms: Conclusions to Dye for," *Neurotherapeutics* 9, no. 3 (July 2012): 599-609, doi:10.1007/s13311-012-0133; Edmund J. S. Sonuga-Barke, Daniel Brandeis, Samuele Cortese, David Daley, Maite Ferrin, Martin Holtmann, Jim Stevenson, et al., "Nonpharmacological Interventions for ADHD: Systematic Review and Meta-Analyses of Randomized Controlled Trials of Dietary and Psychological Treatments," *American Journal of Psychiatry* 170, no. 3 (March 2013): 275-89, doi:10.1176/appi.ajp.2012.12070991; L. Eugene Arnold, Elizabeth Hurt, and Nicholas Lofthouse, "Attention-Deficit/Hyperactivity Disorder: Dietary and Nutritional Treatments," *Child and Adolescent Psychiatric Clinics* 22, no. 3 (May 2013): 381-402, doi:10.1016/j.chc.2013.03.001; Jim Stevenson, Jan Buitelaar, Samuele Cortese, Maite Ferrin, Eric Konofal, Michel Lecendreux, et al., "Research Review: the Role of Diet in the Treatment of Attention-Deficit/Hyperactivity Disorder - an Appraisal of the Evidence on Efficacy and Recommendations on the Design of Future Studies," Journal of Child Psychology and Psychiatry 55, no. 5 (February 2014): 416-27, doi:10.1111/jcpp.12215; Stephen V. Faraone and Kevin M. Antshel, "Towards an Evidence-Based Taxonomy of Nonpharmacologic Treatments for ADHD," *Child and Adolescent Psychiatric Clinics of North America* 23, no. 4 (October 2014): 965-72, doi:10.1016/j.chc.2014.06.003; Joel T. Nigg and Kathleen Holton, "Restriction and Elimination Diets in ADHD Treatment," *Child and Adolescent Psychiatric Clinics of North America* 23, no. 4 (October 2014): 937-53, doi:10.1016/j.chc.2014.05.010.

eliminating all artificial colors, all artificial flavors, all artificial sweeteners, including aspartame, acesulfame K, neotame, saccharin, sucralose, sodium benzoate, butylated hydroxyanisole and butylated hydroxytoluene, carrageenan, monosodium or monopotassium glutamate, and any hydrolyzed, textured, or modified protein.[192] In the European Union, most products containing artificial food dyes are required to have a warning label.**193** Furthermore, as one article noted, "Banned in Norway, Finland and France, Blue #1 and Blue #2 can be found in candy, cereal, drinks and pet food in the U.S."**194** Yet, in US public schools and day programs funded by state and federal governments, there are no bans on food coloring.

Furthermore, another study "revealed that children with urinary [bisphenol A] BPA concentrations at or above the median were more than 5 times more likely to have an ADHD diagnosis than children with levels below the median."[195] When boys were examined alone, the risk of having an ADHD diagnosis was almost 11 times higher, whereas for girls the risk was only 2.8 times higher. One article noted, "BPA is in many types of plastics and the epoxy resins that line most aluminum cans, as well as thermal papers like receipts."[196] At the same time, many of the individuals in schools, day programs for older individuals, and residences are already on psychotropic medications

192 Nigg and Holton, "Restriction and Elimination Diets in ADHD Treatment," 937-53, appendix A.

193 Laurel Curran, "EU Places Warning Labels on Foods Containing Dyes," *Food Safety News*, July 21, 2010, http://www.foodsafetynews.com/2010/07/eu-places-warning-abels-on-foods-containing-dyes/.

194 Susanna Kim, "11 Food Ingredients Banned Outside the U.S. That We Eat," *ABC News*, June 26, 2013,
http://abcnews.go.com/Lifestyle/Food/11-foods-banned-us/story?id=19457237.

195 Dee Rapposelli, "BPA Plastics Linked to ADHD in Boys," *Psychiatric Times*, July 21, 2016, http://www.psychiatrictimes.com/adhd/bpa-plastics-linked-adhd-boys?GUID=BECFC4AD-C9E5-49F1-9A1B-31D55ADAC41F&rememberme=1&ts=21072016; Shruti Tewar, Peggy Auinger, Joseph M. Braun, Joseph M. Braun, Bruce Lanphear, Kimberly Yolton, Jeffery N. Epstein, Shelley Ehrlich, et al., "Association of Bisphenol A exposure and Attention-Deficit/ Hyperactivity Disorder in a National Sample of U.S. Children," Environmental Research 150 (October 2016): 112-118, doi:10.1016/j.envres.2016.05.040.

196 Paul Taylor/Corbis, "BPA Is Fine, if You Ignore Most Studies about It," *Newsweek*, March 4, 2015, http://www.newsweek.com/2015/03/13/bpa-fine-if-you-ignore-most-studies-about-it-311203.html.

that increase the risks for chronic illnesses, which will be further exacerbated by the consumption of junk food and drinks or from the containers.

Other research has shown that "[s]tates that receive the most sunshine generally have the lowest rates of attention-deficit hyperactivity disorder," and researchers have also suggested "that youngsters' extensive use of modern media such as cell phones and various types of computers might play a role in sunlight's ability to prevent ADHD" and noted "that the use of such electronic devices by youngsters shortly before bedtime has been found to delay sleep onset, shorten sleep duration, and suppress melatonin."[197] Therefore, using electronic devices in the evening may cause ADHD symptoms.

Three studies found the youngest children in classes are more likely to be diagnosed with ADHD than the oldest children. In other words, children who are naturally less mature are diagnosed more often with ADHD, which questions the validity of the diagnosis itself.

*USA Today* reported that according to two studies "[n]early one million children may have been misdiagnosed with attention-deficit hyperactivity disorder, or ADHD, not because they have real behavior problems, but because they're the youngest kids in their kindergarten class…Kids who are the youngest in their grades are 60% more likely to be diagnosed with ADHD than the oldest children."[198] The Michigan State University study included 18,644 kindergarteners from more than one thousand kindergarten programs in the fall of the 1998-1999 school year from different states, who were followed as they grew older.[199] The study also showed that teachers were more likely to report ADHD symptoms for younger students.[200] Furthermore, the study

---

197 Joan Arehart-Treichel, "Sunbaked Regions Often Show Lower ADHD Prevalence," *Psychiatric News*, May 17, 2013, http://psychnews.psychiatryonline.org/doi/10.1176/appi.pn.2013.5a3; Martijn Arns, Kristiaan B. van der Heijden, L. Eugene Arnold, and J. Leon Kenemans, "Geographic Variation in the Prevalence of Attention-Deficit/Hyperactivity Disorder: The Sunny Perspective," *Biological Psychiatry* 74, no. 8 (October 15, 2013): 585-90, doi:10.1016/j.biopsych.2013.02.010.

198 Liz Szabo, "Youngest in Class Get ADHD Label," *USA Today*, August 17, 2010, http://usatoday30.usatoday.com/news/health/2010-08-17-1Aadhd17_ST_N.htm.

199 Todd E. Elder, "The Importance of Relative Standards in ADHD Diagnoses: Evidence Based on Exact Birth Dates," *Journal of Health Economics* 29, no. 5 (September, 2010): 641-56, doi:10.1016/j.jhealeco.2010.06.003.

200 Ibid., 641.

reported that "the youngest children in fifth and eighth grades are nearly twice as likely as their older classmates to regularly use stimulants prescribed to treat ADHD."[201]

Another study had similar results.[202] "About 4.5 million children have been diagnosed with ADHD, according to the studies," *USA Today* notes.[203] The findings showing increased diagnoses of and medication treatment for ADHD by Michigan State University were later confirmed again in another study of 937,943 children done in Canada.[204]

Another study contradicted these findings, although the new study used only 1,091 subjects,[205] and the principal author, Dr. Joseph Biederman, was previously sanctioned by Harvard Medical School and Massachusetts General Hospital "for violating conflict of interest policies...to report industry-sponsored activities to the University."[206] Despite the variations in prevalence, there are true cases of ADHD. Nevertheless, a noninvasive solution to decrease the

---

201 Ibid.

202 William N. Evans, Melinda S. Morrill, Stephen T. Parente, "Measuring Inappropriate Medical Diagnosis and Treatment in Survey Data: The Case of ADHD among School-Age Children," *Journal of Health Economics* 29, no. 5 (September 2010): 657-73, doi:10.1016/j.jhealeco.2010.07.005.

203 Szabo, "Youngest in Class Get ADHD Label," *USA Today*, August 17, 2010, http://usatoday30.usatoday.com/news/health/2010-08-17-1Aadhd17_ST_N.htm.

204 Richard L. Morrow, Jane Garland, James M. Wright, Malcolm Maclure, Suzanne Taylor, and Colin R. Dormuth, "Influence of Relative Age on Diagnosis and Treatment of Attention-Deficit/Hyperactivity Disorder in Children," *Canadian Medical Association Journal* 184, no. 7 (April 17, 2012): 755-62, doi:10.1503/cmaj.111619: "Boys who were born in December were 30% more likely (relative risk [RR] 1.30, 95% confidence interval [CI] 1.23-1.37) to receive a diagnosis of ADHD than boys born in January. Girls born in December were 70% more likely (RR 1.70, 95% CI 1.53-1.88) to receive a diagnosis of ADHD than girls born in January. Similarly, boys were 41% more likely (RR 1.41, 95% CI 1.33-1.50) and girls 77% more likely (RR 1.77, 95% CI 1.57-2.00) to be given a prescription for a medication to treat ADHD if they were born in December than if they were born in January" (p. 755).

205 Joseph Biederman, Carter R. Petty, Ronna Fried, K. Yvonne Woodworth, and Stephen V. Faraone, "Is the Diagnosis of ADHD Influenced by Time of Entry to School? An Examination of Clinical, Familial, and Functional Correlates in Children at Early and Late Entry Points," *Journal of Attention Disorders* 18, no. 3 (April 2014): 179-85, doi:10.1177/1087054712445061.

206 Xi Yu, "Three Harvard Professors Face Sanctions Following Harvard Medical School Inquiry, Investigation by Medical School and Massachusetts General Hospital Punishes Psychiatrists Accused by Senator," *Harvard Crimson*, July 2, 2011, http://www.thecrimson.com/article/2011/7/2/school-medical-harvard-investigation/.

incidence of ADHD would be to extend the school day and intersperse recess with moderate-intensity aerobic exercise or other activities, such as playing an instrument, both of which, research has shown, can help with attention skills.[207] Interspersing exercise, recess, music, and art in between academic periods to increase attention has been done in Japan. In fact, in both Japan and Finland, students get breaks of ten to fifteen minutes every hour at school,[208] and according to a research article regarding ADHD prevalence, "Japanese and Finnish children scored lowest."[209] Of course, there are other differences between the Japanese, Finnish, and American environments besides the schedules of school days, and genetic differences as well as differences in diagnosis may contribute to variations in prevalence. However, in the United States, we have eliminated music, art, and recess in many schools, and we just medicate.

Furthermore, as pointed out in one study including 24,372 children and adolescents in eight state Medicaid programs, individuals are receiving antipsychotics rather than behavioral or psychotherapy even when psychosocial approaches are supposed to be first-line treatment for their diagnoses. "For many children and adolescents with ADHD and disruptive disorders, it is likely that aggression and disruptive behaviors are the clinical targets of antipsychotic treatment. Clinical practice guidelines addressing the management of aggression or disruptive disorders in children and adolescents recommend the use of psychosocial interventions before use of antipsychotics, including

---

207 Karen Dineen Wagner, "Mental Health Benefits of Exercise in Children," *Psychiatric Times*, January 2015, 37-38, http://www.psychiatrictimes.com/child-adolescent-psychiatry/mental-health-benefits-exercise-children; Matthew B. Pontifex, Brian J. Saliba, Lauren B. Raine, Daniel L. Picchietti, and Charles H. Hillman, "Exercise Improves Behavioral, Neurocognitive, and Scholastic Performance in Children with Attention-Deficit/Hyperactivity Disorder," *Journal of Pediatrics* 162, no. 3 (March 2013): 543-51, doi:10.1016/j.jpeds.2012.08.036; Ines Jentzsch, Anahit Mkrtchian, and Nayantara Kansal, "Improved Effectiveness of Performance Monitoring in Amateur Instrumental Musicians," *Neuropsychologia* 52, no. 100 (January 2014): 117-24, doi:10.1016/j.neuropsychologia.2013.09.025.

208 Emily Sohn, "Recess: It's Important. Does Your Child Get Enough of It?" *Washington Post*, November 9, 2015, https://www.washingtonpost.com/national/health-science/recess-its-important-does-your-child-get-enough-of-it/2015/11/09/ab610866-8180-11e5-9afb-0c971f713d0c_story.html.

209 Terrie E. Moffitt and Maria Melchior, "Why Does the Worldwide Prevalence of Childhood Attention Deficit Hyperactivity Disorder Matter?" *American Journal of Psychiatry* 146, no. 6 (June 2007): 857, doi:10.1176/appi.ajp.164.6.856.

children with autism. In the present study, children and adolescents with autism and pervasive developmental disorders or ADHD were less likely to receive a psychosocial intervention before receiving antipsychotics than were youth with psychotic or BD [bipolar disorder] diagnoses."[210] "The results of this study suggest that psychotherapy and behavioral interventions are not consistently used as first-line treatment options for Medicaid-financed youth before starting antipsychotic treatment. Although many children and adolescents receiving antipsychotics have diagnoses for which psychotherapy is a first-line treatment, only a minority receive psychosocial services before initiating antipsychotic medications."[211] Furthermore, in this study, Medicaid-managed care patients were less likely to receive psychosocial interventions than patients enrolled in straight Medicaid.[212]

In the United States, medication coercion is also used by school systems to manage behaviors. I have had school personnel call me complaining of my patients' behaviors. One public school staffer had the audacity to tell me, "I don't know why this mother is so against medication." I usually suggest officials write a letter to the CSE to tell them that the current placement is not appropriate because they do not have the resources to manage the child's behavior problems. One public school's autism program dean responded, "We don't do that here."

At another public school's autism program, the coordinator told me no other programs could handle the behavior. I then told her there were other types of programs and that my daughter attended a school with a 4:1:3 staffing, as an example. She then told me that because I was a doctor, they listened to me. I did not explain that the people handling my case at the CSE had no idea that I was a doctor.

210 Molly Finnerty, Sheree Neese-Todd, Riti Pritam, Emily Leckman-Westin, Scott Bilder, Sepheen C. Byron, Sarah Hudson Scholle, et al., "Access to Psychosocial Services Prior to Starting Antipsychotic Treatment among Medicaid-Insured Youth," Journal of the American Academy of Child & Adolescent Psychiatry 55, no. 1 (January 2016): 73, doi:10.1016/j.jaac.2015.09.020.

211 Finnerty, Neese-Todd, Pritam, Leckman-Westin, Bilder, Byron, Hudson Scholle, et al., "Access to Psychosocial Services Prior to Starting Antipsychotic Treatment among Medicaid-Insured Youth," 74.

212 Ibid., 73.

I have had patients transferred to other schools, public as well as private, after their former placements requested medication or suspended them. Sometimes, the new schools never complain about their behavior. Once when I was giving a lecture on advocacy, I stated that a parent's consent was needed to allow children to be given medication. One person in the audience, an employee at the Administration for Children's Services (ACS), stated that sometimes school staffers accused parents of neglect for refusing to give their children medication for behavior. She added that many parents did not understand it was their right to refuse for their children to be given medications.

At one school that the DOE had recommended I visit for Talia, the psychologist told me there was an aggressive student whose parent refused to administer medication, so the school called the ACS. It is illegal for medication to be part of an IEP. With the threat of the ACS and Child Protective Services (CPS), school systems try to intimidate parents to place their children on potentially toxic medications. Sometimes medication does not work to manage behavior, and there have been deaths at these positive-only programs, including the two former JRC students after being transferred to other facilities, discussed earlier in this book.

**"Omnibus Budget Reconciliation Act of 1987"**

Unlike in residential settings for people with developmental disabilities, medication is regulated in nursing homes by federal law, as described in the following excerpts:

> On December 22, 1987, Congress enacted comprehensive nursing home reform with the Omnibus Budget Reconciliation Act (OBRA) of 1987 (P.L. 100-203). As part of OBRA 1987, Congress passed the comprehensive Nursing Home Reform Act (NHRA)…NHRA limits the use of "as needed" (P.R.N.) orders and requires efforts to withdraw

> the drug or decrease dosage be made for residents who are receiving psychotropic drugs.[213]

For individuals with developmental disabilities, there is no regulation that mandates efforts to withdraw or decrease the dosage of psychotropic medication. Further regulations in nursing homes also minimize the use of potentially toxic drugs:

> **The Centers for Medicare and Medicaid Services Regulations**
> To implement the requirements of OBRA 1987, on February 2, 1989, the Center for Medicare & Medicaid Services (CMS)...finalized regulations (42 C.F.R §483.25(1)) on anti-psychotic medications and unnecessary drugs. These regulations were further refined in 1991. The CMS regulations state that each resident's drug regimen must be free from unnecessary drugs and define what is considered an unnecessary drug. An unnecessary drug is any drug used:
>
> - in excessive dose;
> - for excessive duration;
> - without adequate monitoring or without adequate indications for its use;
> - in the presence of adverse consequences, which indicate the dosage should be reduced or discontinued;
> - or without specific target symptoms.[214]

Individuals with developmental disabilities have no legal rights and no such protections from unnecessary drugs including antipsychotic medications:

---

213 Janet Rehnquist, Inspector General, *Psychotropic Drug Use in Nursing Homes*, Department of Health and Human Services Office of Inspector General, OEI-02-00-00490, November 2001, 2, https://oig.hhs.gov/oei/reports/oei-02-00-00490.pdf.

214 Ibid., 3.

> In addition, the CMS regulations state that residents who have not previously used antipsychotic drugs should not be given these drugs unless anti-psychotic drug therapy is necessary to treat a specific condition as diagnosed and documented in the clinical record. Further, the regulation mandates that residents who use antipsychotic drugs receive gradual dose reductions and behavioral interventions, unless clinically contraindicated, in an effort to discontinue these drugs.[215]

No federal law mandates behavioral interventions for individuals over the age of twenty-one with developmental disabilities. Furthermore, there are no federal investigative protocols or any federal program to monitor drug use or quality of care for these individuals as there is for nursing homes, as described below:

> In July 1995, the CMS completed the final step in the implementation of OBRA 1987 by introducing new tasks into the nursing home survey process. Further refinements to survey procedures and interpretive guidelines were implemented on July 1, 1999. Some of these are specific to psychotropic drug therapy, including new investigative protocols and the incorporation of quality indicators (QIs) based on the Minimum Data Set (MDS) into the off-site survey process; an update to the list of anti-psychotic drugs that can be cited as unnecessary or misused under tag number F329; the addition of drug therapy guidelines; and a revised definition of medication error. Additionally, the updated CMS regulations change some anti-psychotic dosing restrictions and list medications considered potentially hazardous to the elderly.[216]

**Nursing Home Monitoring Programs**

Several programs exist to monitor the quality of care in nursing homes. These programs can act as a safeguard against the use of psychotropic

215 Ibid.

216 Ibid.

> drugs as inappropriate chemical restraints and provide a venue for receiving and handling complaints about such misuse.[217]

If such safeguards were for individuals with developmental disabilities, there would be more accountability for the quality of care. Unannounced inspections would help.

> **Nursing Home Survey and Certification**
> All nursing homes are subject to an unannounced standard survey... a surveyor may cite the nursing home for a deficiency if that facility fails to meet specific requirements...Survey deficiencies are categorized, collected, and reported in the Online Survey Certification and Reporting System...Several deficiencies relate directly to psychotropic drug misuse. For example, tag number F329 is cited for all unnecessary drugs including: anti-anxiety, anti-depressant, anti-psychotic, and hypnotic drug use. Tag number F330 is specific to anti-psychotic drug use in residents who do not have a specific condition as diagnosed and documented in the medical record, and tag number F222 is cited for inappropriate chemical restraints. In addition, other tags that address quality of care, resident's rights, and drug regimen reviews also indirectly relate to the misuse of psychotropic drugs.[218]

Unlike in nursing homes, there is no federal law that mandates unannounced inspections for individuals with developmental disabilities who reside in residential settings.

To further enhance reduction in antipsychotics in nursing homes for dementia, a new program was started in 2012, as noted by the American Psychiatric Association:

> Use of antipsychotics in nursing homes for dementia and other conditions has fallen since the start of a new government program,

---

217 Ibid., 3-4.

218 Ibid., 4.

> according to data from the Centers for Medicare and Medicaid Services (CMS)...[before] more than 17% of nursing home patients had daily dosages of antipsychotics exceeding recommended levels... The new data show the national prevalence of antipsychotic use in long-stay nursing home residents had been reduced by 9.1% by the first quarter of 2013, compared with the last quarter of 2011.[219]

While I have seen individuals have improved functioning with psychotropic medication, all too often it is the opposite. The side effects destroy their quality of life. We must stop chemically as well as physically locking up people. We need to do what is right, which is for these individuals to function to the best of their ability and to have a better quality of life.

## Alternative Treatments: Limitations and Risks

The same individuals and groups who decry the dangers of vaccines are often avid proponents of unsupervised untested diets and mega doses of vitamins and other supplements, for which the dangers are sometimes much more evident than the risk of any vaccine. It is clear that without vaccines, people develop illnesses that cause permanent brain damage, profound intellectual disability, and death.

Regarding gluten- and casein-free diets, the study I discussed earlier when I described my concern about Talia having a protein deficiency involved thirty-six children with autism, ten of whom were on the gluten- and casein-free diet. This study demonstrated that children on the diet had an increased prevalence of essential amino acid deficiencies and lower plasma levels of essential acids, including the neurotransmitter precursors tyrosine

219 "Antipsychotic Use Drops in Nursing Homes, But 2013 Goal Not Yet Reached," *American Psychiatric Association Psychiatric News Alert*, August 27, 2013, http://alert.psychnews.org/2013/08/antipsychotic-use-drops-in-nursing.html; Centers for Medicare and Medicaid Services, "New Data Show Antipsychotic Drug Use Is Down in Nursing Homes Nationwide," *Lund Report*, August 27, 2013, https://www.thelundreport.org/content/new-data-show-antipsychotic-drug-use-down-nursing-homes-nationwide.

and tryptophan.[220] The tryptophan deficiency is especially problematic, given that decreased serotonin synthesis has been found in the thalamic and frontal areas of the brains of boys with autism who were not on restricted diets.[221] Without adequate tryptophan to make serotonin, there may be an even worse serotonin deficiency. This diet must therefore be medically supervised. Also, applied behavior analysis-the autism treatment with the most research evidence-often relies on edible reinforcers; the diet can significantly limit the use of edibles, which may impair progress.

I have met parents who have claimed benefits from the diet, but this could possibly be a Hawthorne effect. "The Hawthorne effect refers to a phenomenon where a study subject's behavior and/or study outcomes are altered as a result of the subject's awareness of being under observation."[222] A parent or therapist may relate to the child differently on the diet having greater expectations or providing more attention. Overall, "[c]urrent evidence for efficacy of these diets is poor,"[223] and the "[e]limination diet does not appear effective in treating ASD core symptoms: the fact that individual patients may benefit from special dietary interventions could be hypothesized as the result of subclinical intolerance to specific food allergen."[224]

Some treatments can be downright toxic. For example, some alternative treatment practitioners have recommended a vitamin B-6 intake of up to 500

---

220 Arnold, Hyman, Mooney, and Kirby, "Plasma Amino Acids Profiles in Children with Autism: Potential Risk of Nutritional Deficiencies," 449-54.

221 Chugani, Muzik, Rothermel, Behen, Chakraborty, Mangner, da Silva, et al., "Altered Serotonin Synthesis in the Dentatothalamocortical Pathway in Autistic Boys," 666-69.

222 Rita Mangione-Smith, Marc N. Elliott, Laurie McDonald, Elizabeth A. McGlynn, "An Observational Study of Antibiotic Prescribing Behavior and the Hawthorne Effect," *Health Services Research* 37, no. 6 (December 2002): 1604, doi:10.1111/1475-6773.10482.

223 Claire Millward, Michael Ferriter, Sarah J. Calver, and Graham G. Connell-Jones, "Gluten- and Casein-Free Diets for Autistic Spectrum Disorder." Cochrane Database of Systematic Reviews 2, CD003498 (April 16, 2008), doi:10.1002/14651858.CD003498.pub3.

224 Natascia Brondino, Laura Fusar-Poli, Matteo Rocchetti, Umberto Provenzani, Francesco Barale, and Pierluigi Politi, "Complementary and Alternative Therapies for Autism Spectrum Disorder," *Evidence-Based Complementary and Alternative Medicine* 2015, 258589 (June 2015): 26, doi:10.1155/2015/258589.

or 600 mg daily along with magnesium.[225] However, any dose of vitamin B-6 above 100 mg daily or prolonged use may cause peripheral neuropathy, and the doses of magnesium can cause heart problems among other adverse effects.[226]

## Painful Situations, Bullying, and Murder

In a large classroom with poor structure, some individuals who cannot tolerate the situation may yell or scream, which can be painful to peers who cannot tolerate the noise. The same state government that prohibits aversives has no problems with individuals being exposed to a painful environment in this fashion. As ineffective positive-behavior supports and medication cannot necessarily control aggression, individuals in day programs and residences have to live with and even share their rooms with aggressive peers whom they come to fear because they never know if the roommate will hit them; the state does not recognize a right for these people to have freedom from fear. It is difficult to imagine what it would be like to share a bedroom with someone you fear.

Inadequate funding results in a lack of sufficient staff, training, and supervision. This in turn causes an inability to have adequately intensive positive-behavior supports due to a lack of effective behavior plans with poor consistency. The regulations discussed earlier further constrain the methodologies permitted in a behavior plan, even if other methods used are ineffective. The system thereby maintains verbal and physical abusers. One published article reported the following murder at a group home:

> Police have made an arrest in the death of a man at a home for mentally disabled adults in Downtown Brooklyn. Police say 45-year-old David Baron was found with head trauma Sunday morning at the

---

225 Bernard Rimland, "Vitamin B6 (and Magnesium) in the Treatment of Autism," *Autism Research Review International* 1, no. 4 (1987), http://www.bridges4kids.org/news/9-02/ARI9-02.html.

226 Michelle Zimmer and Cynthia A. Molloy, "Complementary and Alternative Therapies for Autism," in *Clinical Manual for the Treatment of Autism*, ed. Eric Hollander and Evdokia Anagnostou (Washington, DC: American Psychiatric Publishing, 2007), 275.

> Institute for Community Living on Nevins Street. He was taken to Long Island College Hospital where he died. Another resident at the home, 32-year-old Leonardo Martinez, has been charged with second-degree murder, according to investigators.[227]

Despite this death, the OPWDD has recommended this agency serve some JRC students that the OPWDD wanted to return to New York.

Less dramatic episodes of bullying are commonplace due to the lack of effective behavior plans. The nonaversive movement has resulted in "policies that over time can and have produced individuals who become increasingly self-injurious, routinely damage their homes or classrooms, or intimidate and injure peers and staff."[228]

Bullying can produce depression and anxiety. A study of children aged eleven to thirteen years found that "[b]ully-victims may be the most impaired subtype with respect to depression and anxiety."[229] Another study found that "children who were bullied were found to be around five times more likely to experience anxiety and nearly twice as likely to report depression and self-harm at age 18 than children who were maltreated...(defined as physical, emotional, or sexual abuse, or severe maladaptive parenting, or both)."[230] In someone with a developmental disability, depression and anxiety may manifest itself

---

227 "Leonardo Martinez Arrested in Death of David Baron at Institute for Community Living," *Brooklyn News*, April 23, 2013, http://brooklyn.news12.com/news/leonardo-martinez-arrested-in-death-of-david-baron-at-institute-for-community-living-1.5125064?firstfree=yes.

228 Newsom and Kroeger, *Controversial Therapies for Developmental Disabilities: Fad, Fashion, and Science in Professional Practice*, 415.

229 Susan M. Swearer, Samuel Y. Song, Paulette Tam Cary, John W. Eagle, and William T. Mickelson, "Psychosocial Correlates in Bullying and Victimization, the Relationship Between Depression, Anxiety and Bully/Victim Status," *Journal of Emotional Abuse* 2, no. 2-3 (2001): 95-121, summary, doi:10.1300/J135v02n02_07.

230 "Study Finds Victims of Childhood Bullying Are More Likely to Have Mental Health Problems Than Those Maltreated," *American Psychiatric Association Psychiatric News Alert*, April 30, 2015, http://alert.psychnews.org/2015/04/study-finds-victims-of-childhood.html; Suzet Tanya Lereya, William E. Copeland, E. Jane Costello, and Dieter Wolke, "Adult Mental Health Consequences of Peer Bullying and Maltreatment in Childhood: Two Cohorts in Two Countries," *Lancet Psychiatry* 2, no. 6 (June 2015): 524-31, doi:10.1016/S2215-0366(15)00165-0.

as agitation and dangerous behavior,[231] which results in more psychotropic medications with more dangerous side effects.

Being a victim of bullying can have long-term effects. The authors of one study stated, "Bully-victims were much more likely to fail to complete their education, have difficulty holding a job, and have problems forming and maintaining friendships."[232]

In our society, it is unacceptable to have children in an environment where they will randomly be hit, kicked, have some object thrown at them, or otherwise be terrorized. There is concerted effort now to stop bullying at typical schools. This raises the question of why is bullying not being addressed for individuals with developmental disabilities. The disabled individuals in residences, unlike at least some schoolchildren, cannot even go to a safe home at the end of the day.

## Abuse

Physical and sexual abuse takes place in these environments. A number of my patients have been physically and sexually abused at public and private schools, as well as at residences and day programs for older individuals. One colleague told me how individuals with intellectual disabilities came to his office with opportunistic infections, only to find that they were HIV positive and that no one knew how they contracted the virus. As one article stated:

> In hundreds of cases reviewed...employees who sexually abused, beat or taunted residents were rarely fired, even after repeated offenses,

231 Ozinci, Kahn, and Antar, "Depression in Patients with Autism Spectrum Disorder," 294; Joshua Nadeau, Michael L. Sulkowski, Danielle Ung, Jeffrey J. Wood, Adam B. Lewin, Tanya K. Murphy, Jill Ehrenreich, et al., "Treatment of Comorbid Anxiety and Autism Spectrum Disorders," *Neuropsychiatry* 1, no. 6 (December 2011): 568, doi:10.2217/npy.11.62.

232 "Child Bully Victims Fare Worse as Adults on Several Measures, Study Finds," *American Psychiatric Association Psychiatric News Alert*, August 21, 2013, http://alert.psychnews.org/2013/08/child-bully-victims-fare-worse-as.html; Dieter Wolke, William E. Copeland, Adrian Angold, and E. Jane Costello, "Impact of Bullying in Childhood on Adult Health, Wealth, Crime, and Social Outcomes," *Psychological Science* 24, no. 10 (October 2013): 1958-70, doi:10.1177/0956797613481608.

> and in many cases, were simply transferred to other group homes run by the state...State records show that of some 13,000 allegations of abuse in 2009 within state-operated and licensed homes, fewer than 5 percent were referred to law enforcement...The state has no educational requirements for the positions, which involve duties like administering drugs, driving residents to day activities, feeding them and preventing them from choking...[A]n employee at a group home in Utica, was convicted of beating a 99-year-old man while moonlighting at a nursing home-slapping the man three times in the face and once on the stomach...But he kept his state job working with the developmentally disabled...[A]fter they [workers] arrive at their new workplaces, they often abuse again.[233]

When symptoms develop of posttraumatic stress disorder, it is difficult, if not impossible, to use psychotherapy to treat individuals with communication impairments.

The OPWDD will not allow cameras in their funded facilities, supposedly to protect individuals' privacy. At Stuart's agency, OPWDD demanded cameras be removed from a common area at a residence. However, what really is protected is the privacy of the staff to abuse the patients. Staffers in New York residences are not required to have urine drug screens, although in many residences only one staffer is awake at night, and therefore no one could detect if that staffer is working under the influence of a drug. Many patients have language limitations, so without proof from a camera, they cannot testify, and the perpetrators remain employed.

One of my patients suffered horrible sexual abuse at a school the DOE wanted us to consider for Talia. When these cases occur at schools, parents tell me their children cannot testify, so the staffers involved never have to leave. I sometimes wonder if working with this population attracts deviants who

---

233 Danny Hakim, "At State Run Homes, Abuse and Impunity," *New York Times,* March 12, 2011,
http://www.nytimes.com/2011/03/13/nyregion/13homes.html?_r=1.

know they can get away with abusing a population that is not competent to testify. Without cameras, there can be no justice.

Sometimes it may just be frustrated, overwhelmed staff members who have not been given the needed support, training, and supervision and may give in to impulsive actions. Underfunding and understaffing are a breeding ground for abuse. Furthermore, because of underfunding, poorly implemented positive-behavioral interventions (further limited by the policy of not incurring "human rights violations" discussed elsewhere), and aversive prohibitions, staffers have to deal with dangerous behaviors, and sometimes they may end up in the hospital themselves.

At one facility at which I worked, I found out that one staff person witnessed abuse by a colleague but was pressured by other staff not to "snitch." In a different case, a newspaper reported "[f]amily members say he was branded with a potato masher...The Bronx District Attorney's Office says it did investigate the incident but 'could not meet the burden of proof beyond a reasonable doubt because of the victim's medical condition.'" According to the news report, the individual could not communicate what was done to him.[234] It is difficult to convict someone of any crime when the victim does not have the communication skills to testify.

In New York, a law was passed in 2007, known as "Jonathan's Law," to stop abuse.

> An amendment to Jonathan's law, which allowed parents to obtain retroactive records dating back to January 2003, was signed by the governor...However, the amendment expired on Dec. 31, 2007. According to Gary Masline, assistant council and special assistant to the chair of the Commission on Quality of Care and Advocacy for Persons with Disabilities, an announcement was posted on the organization's Web site informing people of the amendment on Oct. 30, 2007.[235]

---

234 "Family of Autistic Burn Victim Eduardo Sandobal Files Civil Suit," *News 12 The Bronx*, May 20, 2013, http://bronx.news12.com/family-of-autistic-burn-victim-eduardo-sandobal-files-civil-suit-1.5302771.

235 Maria Brandecker, "Jonathan's Law Advocates See Need for Further Reform," *Legislative Gazette* 31, no. 25 (February 25, 2008): 24, http://www.legislativegazette.com/PDF/08-2-25.pdf.

It makes no sense to allow such an amendment to have an expiration date.

At most placements I visit, parents are told they need to make appointments to observe their children in the facility. This raises questions about what it is these facilities are trying to hide. An open-door policy protects all.

To see my brothers firsthand and my abused patients and patients in fear of abuse by peers made me contemplate moving out of New York State before my daughter turned twenty-one, so she could be a resident of another state. My parents have considered moving to Massachusetts to get Stuart admitted to the JRC. My mother consulted with an attorney, only to find out that if she moved, my brother would not be able to get admission there and he would still be a resident of New York because he was residentially placed in New York and was older than twenty-one.

## New York State Education Department and Aversives

I find it ironic that New York has a firm anti-aversive stance. A New York State Education Department report on the JRC made numerous false statements including that "[t]he education program is organized around the elimination of problem behaviors largely through punishment,"[236] even though the program has always been mostly positive reinforcement. The report has also stated that JRC does not routinely utilize preference assessments to determine reinforcers,[237] although the school has shared their preference assessment forms with me, which I utilized at one of my jobs. In fact, prior to my brother's entry to the school, I recall filling out pages of forms of preference assessments with my mother. The New York State Education Department sent three psychologists to investigate the JRC for that report, one of whom was allegedly misidentified as a "Regional Associate." The New York State Education Department also reported suicidality as a side effect.[238] Their report specifically stated that a student was suicidal. At the time of the investigation,

---

236 "Observations and Findings of Out-of-State Program Visitation Judge Rotenberg Educational Center," *New York State Education Department JRC Program Visitation Report*, June 9, 2006, 17.

237 Ibid., 18.

238 Ibid., 26.

the psychologists never alerted the staff or family that this student was suicidal. If this student was suicidal, as they had claimed, not taking safety measures would be unprofessional, unethical, and constitute malpractice.

Deputy Commissioner Rebecca Cort supported anti-aversive regulations. In January 2007, she asked the Board of Regents if they wanted these decisions to be decided by a court (instead of them). A court decision via the impartial hearing process, unlike adherence to generalized regulations, would be individualized to the child and more impartial. Richard Mills, the commissioner of the New York State Education Department, urged the Board of Regents to pass the anti-aversive regulations on which they were about to vote. He said he wanted them to support Rebecca Cort. Therefore, the Board of Regents was told not to vote on the merits of the regulations but rather what appears to be a decision based on political reasons.

This prohibition on aversives continued through federal court for individuals who had not received them previously. The federal court grandfathered students who were already receiving them. On August 20, 2012, the regulation was upheld in a 2:1 decision on appeal in a summary judgment. The dissenting opinion of Justice Dennis Jacobs noted the following:

> It is worth noting that of the several studies cited in the Notice of Emergency Adoption and Proposed Rule Making, the two included in full in the record actually describe the need for aversive interventions in certain instances. See Dorothy C. Lerman & Christina M. Vondram, On the Status of Knowledge for Using Punishment: Implications for Behavior Disorders, 35 J. Applied Behavior Analysis, 431, 456 (2002) (noting that "punishment is still sometimes needed to reduce destructive behavior to acceptable levels"); Sarah-Jeanne Salvy et al., Contingent Electric Shock (SIBIS) and a Conditioned Punisher Eliminate Severe Head Banging in a Preschool Child, 19 Behavior Interventions, 59, 70 (2004) (noting that ABIs[239] "can sometime be necessary, although not sufficient, to eliminate severe

239 SIBIS stands for "self-injurious behavior inhibiting system," a type of skin-shock device. ABI is "aversive behavioral intervention."

> and harmful [self-injurious behavior] in the natural environment"). Consequently, I am unpersuaded that the Notice of Emergency Adoption and Proposed Rule Making cited by the majority provides a sufficient basis for upholding the district court's dismissal.[240]

However, as Talia's case has made clear, they have no problem running programs of poor quality that allow the children's behaviors to escalate into something that may later require the use of aversives.

In March 2011, the New York State Education Department, after previously installing regulations prohibiting the use of aversives, decreased mandated services to children with special needs. Yet, in their JRC report, they accused JRC of not providing related services,[241] even though students at JRC not receiving related services did not have these services on their IEPs.

Prior to March 2011, if a child from New York State required speech therapy, there had to be a minimum of two times a week for thirty minutes each session. Furthermore, a child classified with autism had to have daily language instruction of at least thirty minutes, which some impartial hearing officers interpreted to mean speech *and* language therapy five times a week for thirty minutes each session (not that the school districts were informing parents of this). Denying speech therapy contributes to problems, because improved communication can help decrease problem behavior. "For example, it appears that when communication skills are improved, the rate of inappropriate behaviors decreases."[242]

"Another variable that makes effective instruction difficult is the various forms of inappropriate behavior (e.g., aggression, self-stimulation, withdrawal) that are commonly associated with ASD."[243] This can include dangerous behaviors such as head banging or other life-threatening behaviors, as

---

240 Bryant v. New York State Education Dept., 692 F.3d 221, US Court of Appeals (2d Cir. August 20, 2012), docket 10-4029, dissent, http://caselaw.findlaw.com/us-2nd-circuit/1609202.html.

241 *New York State Education Department JRC Program Visitation Report*, 23.

242 Fry Williams and Lee Williams, *Effective Programs for Treating Autism Spectrum Disorder: Applied Behavior Analysis Models*, 125.

243 Ibid., 147.

well as non-life-threatening behaviors that interfere with learning. Authors Sundberg and Partington have stated that such "behaviors come to function as the child's main form of communication."[244] For example, children may scream or get aggressive to get attention or avoid work because they cannot communicate verbally their needs or desires, such as requesting a break. As another writer noted, "Sundberg and Partington (1998) [*Teaching Language to Children with Autism or Other Developmental Disabilities*] point out that programs that attempt simply to decrease inappropriate behaviors are likely to fail. Language training needs to be a major component of programs for children with ASD."[245] If the self-appointed anti-aversive advocates really cared about people with special needs, they would focus their efforts on the scarcity of early intervention and special education for all children who needed it to decrease the risk of a behavior requiring an aversive in the future.

A social worker at the CSE wrote a letter to the New York State Education Department requesting approval for Talia's 1:1 ABA school, but it did not happen. In fact, in New York there is a clear segregation of special education, because most minority families cannot afford lawyers or private evaluations, much less 1:1 ABA school tuition. This is where advocacy needs to be focused. Any child who needs a 1:1 ABA program should be able to have one, regardless of the parents' socioeconomic status. The fact that a 1:1 ABA charter school exists in New York City, although it teaches only twenty-eight students, shows that the government acknowledges the need for such an environment. While the New York State Education Department focuses on banning aversives, it refuses to approve 1:1 ABA programs to all who need it. In addition, if the district-recommended school fails the child, and the child attends a non-state-approved private school that has an unsuccessful program, the parents are left holding the tuition bag, and it's usually a big bag.

Just as there are staff at the New York State Education Department who think they know what is better for Talia-although they have never met my daughter, read any of her evaluations, or visited her school-there are

244 Mark L. Sundberg and James W. Partington, *Teaching Language to Children with Autism or Other Developmental Disabilities* (Danville, CA: Behavior Analysts, 1998), 5.

245 Fry Williams and Lee Williams, *Effective Programs for Treating Autism Spectrum Disorder: Applied Behavior Analysis Models*, 147.

self-appointed anti-aversive activists who have never met Stuart or Matthew, read any of their evaluations, or visited their placements and yet think they know what is appropriate for them better than my family and I do.

## Anti-aversive Organizations

The use of aversives decreases and even stops the need for psychotropic medications, as it did for my Matthew. If a pharmaceutical company spoke out against aversives, it would appear suspicious and could be counterproductive for the company's objectives, so instead they fund "advocacy" groups to do it for them. I have tried the skin shock to find out firsthand what my brother experiences, but I refuse to try Risperdal and the like, because I know psychotropic drugs could seriously harm or even kill me.

### *JRC 2012 Lawsuit and Autism Speaks/ASA*

I was disgusted with a tape that a judge allowed to be released to the public from a lawsuit by a parent claiming that her child was "tortured" when given multiple skin shocks (the parent had signed a consent form to allow the use of skin shocks). This tape showed a child undergoing skin shocks to a public that, for the most part, had no experience with individuals with life-threatening behaviors. By allowing the tape to go public, the judge deterred any facility working with the developmentally disabled population from video monitoring, which is vitally needed to prevent abuse and neglect.

I was also disgusted with Autism Speaks and the Autism Society of America (ASA) for issuing condemning statements without visiting the school or speaking to any family member or individual who believed such treatment saved their family member's or their own lives.[246] The ASA called the skin-shock

246 Kim Wombles, "Autism Speaks Issues Statement Concerning Judge Rotenberg Center," *Science 2.0*, April 18, 2012, http://www.science20.com/countering_tackling_woo/autism_speaks_issues_statement_concerning_judge_rotenberg_center-89143; Kim Wombles, "Autism Society Issues Statement Regarding the Judge Rotenberg Center," *Science 2.0*, April 19, 2012, http://www.science20.com/countering_tackling_woo/autism_society_issues_statement_regarding_judge_rotenberg_center-89202.

treatment archaic,[247] despite the extensive literature on skin-shock treatments I have discussed in this book.

Reading these condemning statements disturbed my sleep that night. I even wrote a letter to Autism Speaks on how skin shocks saved Matthew's life and how his identical twin engaged in life-threatening behavior and could not get access to appropriate care. I never even received the courtesy of a response.

### *Anti-Aversive Organizations and Agendas*

I began to research Autism Speaks and its connection to the pharmaceutical industry.

According to the organization's website,

> Autism Speaks…has named Robert H. Ring, Ph.D…vice president of translational research…with the goal of improving outcomes for individuals with autism spectrum disorders (ASD).
>
> Since 2009, Dr. Ring has served as senior director, and head of the Autism Research Unit at Pfizer, the first research and development unit in the pharmaceutical industry entirely focused on developing medications for treating people with ASD and related neurodevelopmental disorders.[248]

The panel of professional advisors to the Autism Society includes Gary LaVigna and Anne M. Donnellan,[249] who wrote *Alternatives to Punishment: Solving Behavior Problems with Non-Aversive Strategies.*[250] Critical responses to this book stated, "[T]he cases presented constitute neither the severe

247 Kim Wombles, "Autism Society Issues Statement Regarding the Judge Rotenberg Center," *Science 2.0*, April 19, 2012, http://www.science20.com/countering_tackling_woo/autism_society_issues_statement_regarding_judge_rotenberg_center-89202.

248 "Autism Speaks Names Robert Ring to New Position of Vice President of Translational Research," *Autism Speaks*, May 9, 2011, http://www.autismspeaks.org/about-us/press-releases/autism-speaks-names-robert-ring-new-position-vice-president-translational-re.

249 "Panel of Professional Advisors," *Autism Society*, accessed February 7, 2016, http://www.autism-society.org/about-the-autism-society/boardadvisors/panel-of-professional-advisors/.

250 Gary W. LaVigna and Anne M. Donnellan, *Alternatives to Punishment: Solving Behavior Problems with Non-Aversive Strategies* (New York, NY: Irvington Publishers, 1986).

problem behaviors that practitioners frequently encounter nor do they contain elements of the scientific rigor essential to demonstrating the effectiveness of procedures,"[251] and professionals "who have and do use punishment procedures responsibly would never consider their use with the behaviors that Donnellan and LaVigna treat."[252]

Donnellan has also written in support of facilitated communication, which occurs when a facilitator holds a child's arm or hand while the child types on a keyboard, despite this technique's efficacy being disproven in multiple research studies; innocent parents, falsely accused of abuse on the basis of words typed into the facilitated communication devices, have been arrested or jailed, and families have been torn apart, because of this discredited treatment.[253] Donnellan, together with Paul Haskew, wrote, "Reports that facilitated communicators seem to be able to read their facilitators' and other people's minds surface whenever facilitation is attempted."[254] The suggestion that these people may be reading minds indicates these authors are not credible about anything, including positive-only treatments.

Just as it would be unethical to replace antibiotics (a scientifically effective treatment) with marijuana (an untested treatment) as a medication for pneumonia, it is unethical to replace applied behavior analysis (a scientifically evidence-based effective treatment) with facilitated communication (a hoax). Despite facilitated communication being disproven, TASH, previously known as the Association for Persons with Severe Handicaps, an anti-aversive organization-although admitting facilitated communication to be controversial-supports its use.[255]

---

251 Norman A. Wieseler, "Review of Alternatives to Punishment: Solving Behavior Problems with Non-Aversive Strategies," by G. W. LaVigna and A. M. Donnellan, in *Research in Developmental Disabilities* 9, no. 3 (1988): 323.

252 Foxx, *Controversial Therapies for Developmental Disabilities: Fad, Fashion, and Science in Professional Practice*, 301.

253 Jon Palfreman, "Prisoners of Silence," *PBS Frontline*, October 19, 1993, http://www.pbs.org/wgbh/pages/frontline/programs/trascripts/1202.html.

254 Paul Haskew and Anne M. Donnellan, *Emotional Maturity and Well-Being: Psychological Lessons of Facilitated Communication* (Madison, WI: DRI Press, 1993), 13.

255 TASH Resolution on the Right to Communicate, accessed February 7, 2016, https://tash.org/about/resolutions/tash-resolution-right-communicate/.

TASH's website states the organization is for "equity, opportunity and inclusion." Regarding aversives, TASH has stated the following: "All aversive techniques have in common the application of physically or emotionally painful stimuli."[256] The term "emotionally painful stimuli" is highly subjective. Therefore, saying "no," if it results in crying, can be considered "emotionally painful stimuli," even if the person was being told not to do something dangerous, and would be considered aversive.

Some of the falsehoods I believe TASH asserts regarding aversives include the following: "Aversive and restrictive procedures, including the inappropriate use of restraint, are often used as part of a systematic program for decreasing certain behaviors. They are most often used without the individual's or even a substitute individual's informed consent."[257] This has never been true at the JRC, where the legal guardian must consent to the treatment, and the skin shock must be reviewed and preapproved by a primary care physician, a psychiatrist, a human rights committee, and a court.

"Although it has been believed that such procedures are necessary to control dangerous or disruptive behaviors, it has now been irrefutably proven that a wide range of methods are available that are not only more effective in managing dangerous or disruptive behaviors, but do not inflict pain on, humiliate, dehumanize or overly control or manipulate individuals with disabilities."[258] Research has shown the opposite: the use of nonaversive techniques alone is not always effective, as I have discussed elsewhere in this book. If this statement were true, then there is no reason some individuals (including Matthew) accepted at the JRC would otherwise be stuck in a hospital and be highly medicated as no alternative placement in the United States would accept them because of their dangerous behaviors. So much for "equity, opportunity, and inclusion." I have patients who cannot go into their communities because of their behaviors.

---

256 TASH Letter to Massachusetts Department of Developmental Services, July 26, 2011, http://tash.org/blog/2011/07/26/tash-letter-to-massachusetts-department-of-developmental-services.

257 Ibid.

258 Ibid.

Inclusion also makes no sense in a classroom setting if a student will be looking out the window or watching the ceiling fan. Research does not support inclusion for all children, but inclusion does lower the educational cost for children with special needs.[259]

This is why school districts like "inclusion" so much. When looking short term at the immediate budget, it saves money. In the long term, it wastes money, because the students make no progress.

One thing is clear: exclusive positive-behavior supports can prevent inclusion. For example, one recommended placement for Talia that espoused exclusive positive-behavior supports had a classroom that anyone with glasses was forbidden to enter, because one student would not tolerate it. The child was not being taught to function around others wearing glasses. If someone cannot tolerate others wearing glasses, community inclusion is precluded. Aversives are necessary: "As Sidman (1989) showed convincingly, many behaviors in the natural environment are controlled by aversive contingencies."[260] Nonaversive strategies alone may not prepare an individual for living in the community or work.[261] "Further, research with children indicates an important role for aversives in learning and development."[262]

Furthermore, it is disingenuous for anti-aversive advocates to use the adjective "challenging" instead of "self-injurious" or "aggressive" to describe these behaviors.[263] It is inappropriate and just plain wrong to use euphemisms for life-threatening behavior.

---

259 Devery R. Mock and James M. Kauffman, "The Delusion of Full Inclusion," ed. John W. Jacobson, Richard M. Foxx, and James A. Mulick in *Controversial Therapies for Developmental Disabilities: Fad, Fashion, and Science in Professional Practice,* 113-14.

260 Murray Sidman, *Coercion and Its Fallout* (Boston, MA: Authors Cooperative, 1989); Newsom and Kroeger, *Controversial Therapies for Developmental Disabilities: Fad, Fashion, and Science in Professional Practice,* 417.

261 Newsom and Kroeger, *Controversial Therapies for Developmental Disabilities: Fad, Fashion, and Science in Professional Practice,* 417.

262 Ibid.

263 James A. Mulick and Eric M. Butter, "Positive Behavior Support: A Paternalistic Utopian Delusion," in *Controversial Therapies for Developmental Disabilities: Fad, Fashion, and Science in Professional Practice,* ed. John W. Jacobson, Richard M. Foxx, and James A. Mulick (Mahwah, NJ: Lawrence Erlbaum Associates, 2005), 392.

I am sorry I ever donated money to Autism Speaks or the ASA, and I wish I could take back every penny I ever gave them. In some respects, they have caused more trauma and made me feel more isolated and alienated. They do not focus on the "bread and butter": children's education. So many schools throughout the country are failing these children. Appropriate education is the exception rather than the rule. Individuals over twenty-one have almost no rights to appropriate services. Meanwhile, these organizations say nothing about the excessive use of medication.

I use Autism Speaks via the Autism Genetic Research Exchange to analyze my family's genetics. I agree with their advocacy for licensing board-certified behavior analysts. In addition, Autism Speaks did finance a security system for my home for two years to prevent Talia from running away. However, I do not like ASA and Autism Speaks deceiving the public, Autism Speaks appointing an individual from a pharmaceutical company to their vice president of translational research, these organizations' absence of adequately promoting tested and proven therapies, and their other avoidances of the real issues and solutions for those with autism and their families.

### *Anti-aversive Organizations and Political Influence*

TASH also has several member organizations in its Alliance to Prevent Restraint, Aversive Interventions and Seclusion (APRAIS).[264]

---

264 These include the American Association of People with Disabilities, the ARC of the United States, Association of University Centers on Disabilities, Autism National Committee, Autism Society of America, Autistic Self Advocacy Network, Children and Adults with Attention-Deficit/Hyperactivity Disorder, Council of Parent Attorneys and Advocates, Developmental Disabilities Nurses Association, Disability Rights Education and Defense Fund, Epilepsy Foundation of America, Families Against Restraint and Seclusion, Family Advocacy and Community Training, Gamaliel Foundation, Inclusion BC, Judge David L. Bazelon Center for Mental Health Law, Keep Students Safe, National Alliance on Mental Illness, National Association of Councils on Developmental Disabilities, National Association of State Mental Health Program Directors, National Autism Association, National Council on Independent Living, National Coalition for Mental Health Recovery, National Disability Rights Network, National Down Syndrome Congress, National Down Syndrome Society, National Fragile X Foundation, Parent to Parent USA, Respect ABILITY Law Center, and the Family Alliance to Stop Abuse and Neglect, accessed February 6, 2016,
http://tash.org/advocacy-issues/coalitions-partnerships/aprais/aprais-member-organizations/.

In 2011, one of the member organizations, the ARC of the United States, received $2,031,477 in government grants,[265] and an anonymous donor gave at least $200,000.[266] In 2011, their total revenue was $ 7,790,301,[267] and in 2012, they received $1,365,341 in government grants.[268]

ARC has political influence to get its government grants. Peter Berns, its CEO, was appointed by President Obama:

> In May 2011, President Barack Obama appointed Berns to the President's Committee for People with Intellectual Disabilities, which will provide advice and assistance to President Obama and the Secretary of Health and Human Services on a broad range of topics that impact people with intellectual and developmental disabilities and their families.[269]

ARC's "work includes: (a) legislative advocacy in Congress, (b) executive and regulatory advocacy with the Administration and officials of federal agencies, and (c) legal advocacy through participation in federal court litigation."[270] ARC often testifies before federal and state legislatures and agencies. ARC receives government money, lobbies the government, and is even part of the government.

Even though ARC is politically active, no donors above a certain amount need to be disclosed to monitor for special interests. After the Supreme Court's

265 Combined Financial Statements, the ARC of the United States, the Foundation of the ARC of the United States for the Year Ended December 31, 2011 with Summarized Financial Information for 2010, 6, www.thearc.org/document.doc?id=3833.

266 *The ARC 2011 Annual Report*, 12, http://www.thearc.org/document.doc?id=3836.

267 Combined Financial Statements, the ARC of the United States, the Foundation of the ARC of the United States for the Year Ended December 31, 2011 with Summarized Financial Information for 2010, 6, http://www.thearc.org/document.doc?id=3833.

268 Combined Financial Statements, the ARC of the United States, the Foundation of the ARC of the United States for the Year Ended December 31, 2012 with Summarized Financial Information for 2011, 6, http://www.thearc.org/document.doc?id=4188.

269 The ARC Media Center, accessed February 6, 2016, http://www.thearc.org/who-we-are/media-center/peter-v-berns.

270 The ARC Public Policy and Legal Advocacy, accessed April 10, 2016, http://www.thearc.org/what-we-do/public-policy.

infamous decision in *Citizens United*, organizations can lobby on issues without disclosing their donors.[271]

TASH is also politically active. Carol Quirk, while serving president of the TASH board of directors,[272] was appointed by President Obama to the President's Committee for Individuals with Intellectual Disabilities.[273] TASH, like ARC, also receives grants, and in 2015, received $275,141.37 in grants, but their report does not disclose where the grants come from.[274]

Multiple organizations-including, but not limited to, the ARC of US, TASH, Autism National Committee (AutCom), United Cerebral Palsy (UCP), University of San Diego Autism Institute, Easter Seals, the National Leadership Consortium on Developmental Disabilities at University of Delaware-signed a complaint letter on September 30, 2009, and sent it to the Department of Justice and other agencies.[275] The letter advocated to stop the use of aversives and specifically condemned the JRC. The letter itself is deceptive because it stated that aversive procedures were "unnecessary methods of behavior modification."[276] Published research and my brothers are living proof of the contrary.

In another letter to the FDA to stop skin shocks, signatories included, among others, the Autism National Committee, various state ARCs, Easter Seals, TASH, and the Autism Institute, University of San Diego. This letter stated outright untruths, such as the following: "Aversives-the use of pain as

271 Jason M. Breslow, "In Big Sky Country, a Campaign Finance Fight Reverberates Still," *PBS Frontline*, August 27, 2015, http://www.pbs.org/wgbh/frontline/article/in-big-sky-country-a-campaign-finance-fight-reverberates-still/.

272 "TASH Board President Appointed to President's Committee," TASH, June 7, 2011, https://tash.org/blog/2011/06/07/tash-board-president-appointed-to-presidents-committee/.

273 "President Obama Announces More Key Administration Posts," White House Office of the Press Secretary, May 10, 2011, https://www.whitehouse.gov/the-press-office/2011/05/10/president-obama-announces-more-key-administration-posts-5102011.

274 "TASH 2015 Annual Report," 20, http://tash.org/wp-content/uploads/2016/05/TASH-2015-Annual-Report-Final-1.pdf.

275 "U.S. Department of Justice Opens Investigation on Judge Rotenberg Center," Lbrb Autism News Science and Opinion, February 23, 2010,
http://leftbrainrightbrain.co.uk/2010/02/23/u-s-department-of-justice-opens-investigation-on-judge-rotenberg-cente/.

276 Ibid.

a means of behavior modification-are an inherently unsafe and unsupported type of medical treatment...[T]here is no empirical evidence that suggests the shocks are effective as a form of treatment in addressing these behaviors."[277] In fact, as discussed previously, 119 peer-reviewed articles have supported aversive skin shock.

The JRC has a policy of discontinuing psychotropic medication, and so it is a threat to the billions of dollars that annually go to the pharmaceutical industry and their retailers. Here are some examples of direct ties of anti-aversive groups to the pharmaceutical industry and corporations that sell medications:

- According to the Easter Seals website: "For more than a decade, CVS has been a corporate partner of Easter Seals."[278]
- The Easter Seals "Chairman's Corporate Roundtable: contributing $1million or more" includes "CVS Caremark Cooperation" and the "Corporate Council: contributing $30,000-99,000" includes Pfizer.[279] Total revenue for Easter Seals in 2012 and 2013 was $83,886,203 and $82,182,422, respectively.[280] In 2011 and 2012, respectively, Easter Seals received $278,844,000 and $268,964,000 in government grants.[281]
- In 2014 and 2015, Walmart was one of the sponsors of the TASH conference.[282]

277 "Letter to Food and Drug Administration on the Judge Rotenberg Center," Autistic Self Advocacy Network, February 12, 2013, http://autisticadvocacy.org/2013/02/letter-to-food-and-drug-administration-on-the-judge-rotenberg-center/.

278 Easter Seals National Corporate Sponsors, accessed January 9, 2016, http://www.easterseals.com/who-we-are/partners//.

279 Ibid.

280 Easter Seals Return of Organization Exempt from Income Tax, form 990, 2013, http://www.easterseals.com/shared-components/document-library/easter-seals-inc-fy-2013-990.pdf.

281 "2012-2013 Easter Seals Disability Services Annual Report," 6, http://www.easterseals.com/shared-components/document-library/2013-2013-annual-report.pdf.

282 TASH Sponsors and Exhibitors, March 21, 2014, http://conference.tash.org/sample-sponsor-page/; *TASH 2015 Annual Report*, 7, http://tash.org/wp-content/uploads/2016/05/TASH-2015-Annual-Report-Final-1.pdf.

- UCP announced it was "proud to welcome the following exhibitors to the 2013 UCP Annual Conference…Merz Pharmaceuticals… Pharmacy Alternatives."[283] Merz was also the top "diamond level" sponsor of the 2015 conference.[284]

Regarding UCP's commitment to individuals with disabilities, in 2015, when UCP took over part of another agency in New York City, at one residence, even though the takeover was planned months in advance, two and a half weeks after the takeover, they had installed no phones. Additionally, there was no contact for medical emergencies, such as an abnormal lab value.

Similarly, in 2015, when UCP took over a day program, they reduced the nonunionized supervisors' salaries, even though some had been good workers for years. One hard-working supervisor in particular told me UCP cut her pay from $48,000 to $40,000. When staffers have their pay cut, they cannot be expected to have the same motivation.

The letters discussed earlier were deceptive for other reasons. The signatories all appeared to be all separate organizations, but a later search found connections between some of the signatories. For example, in northeast Indiana, Easter Seals merged with the ARC.[285] In North Carolina and Virginia, Easter Seals merged with UCP.[286] ARC also partnered with the National Leadership Consortium on Developmental Disabilities.[287] Likewise, according to another website, "Anne Donnellan, Ph.D., has had a long time interest in autism and related disorders and is a Professional Advisory Panel member of the Autism Society of America as well as the Autism National Committee and TASH." It

---

283 United Cerebral Palsy 2013 Annual Conference Exhibitors, http://ucp.org/media/events/events-archive/2013conference/exhibitors/.

284 2015 United Cerebral Palsy Annual Conference Sponsors and Exhibitors, http://ucp.org/wp-content/uploads/2015/03/Meow-e1430940708914.jpg.

285 Easter Seals Arc Northeast Indiana, accessed February 7, 2016, http://www.easterseals.com/neindiana/?referrer=https://www.google.com//.

286 Easter Seals UCP North Carolina and Virginia, accessed February 7, 2016, http://www.easterseals.com/NCVA/.

287 University of Delaware: The National Leadership Consortium on Developmental Disabilities, National Leadership Consortium Partners, accessed February 7, 2016, http://www.nlcdd.org/partners.html.

also stated that Anne Donnellan is the "Professor, Director of USD Autism Institute University of San Diego."[288] Nevertheless, she had supported facilitated communication, despite its research being disproved, as discussed earlier.

Staffers at some placements, such as the New England Center for Children (NECC), have spoken out publically against the JRC. However, the center also sent a letter dated October 25, 2001, to the JRC, because it wanted to transfer a student there that the NECC could not manage. A NECC discharge summary of another student dated March 2005 (sent to me from the JRC, where the student was transferred) stated the student "may require alternative interventions than those normally used at NECC, for example, mechanical restraint or contingent aversive stimulation."

## United Nations and Aversives

I was disappointed when the United Nations (UN) became involved with aversives, which is really outside of their mandate. The UN has never visited Matthew's school or spoken with a family member, yet it labeled my brother's treatment as torture and requested the school to be investigated.[289] This is a response that I wrote in July 2012, which was never published.

> **My Brother's Treatment Has Become the United Nation's Israel of the Disabled**
>
> I have identical twin brothers and a daughter with autism. I am also a psychiatrist who completed a fellowship in autism at the Mount Sinai Medical Center in New York. In 1988, one brother banged his head so severely he had to have surgery and intravenous antibiotics. He was hospitalized over five months on a combination of five psychotropic medications, which did not stop his repeated head banging.

288 "Autism as a Movement Difference," Conversations That Matter, Online Conference Center, Broadreach Training, May 20, 2013,
http://conversationsthatmatter.org/presenters/donnellan-anne.

289 Juan E. Mendez, *Report of the Special Rapporteur on Torture and Other Cruel, Inhuman or Degrading Treatment or Punishment*, Human Rights Council, Twenty-second session, agenda item 3, A/HRC/22/53/Addendum4, March 12, 2013, 84-85.

Prior to this hospitalization, he developed [symptoms of] neuroleptic malignant syndrome, a deadly side effect from haloperidol. The Judge Rotenberg Center, a school in Massachusetts, was the only school in the United States that would accept him. The two-second skin-shocks that my parents consented to-a treatment that I tried myself, that was court approved, that is based on over 100 peer reviewed articles, and that saved Matthew's life-are a treatment that was labeled torture by the previous UN special rapporteur on torture, Manfred Nowak. Furthermore, Juan Mendez, the current special rapporteur on torture, as his predecessor, has requested the school be investigated for torture. My brother has been off all psychotropic medications over twenty years, no longer bangs his head, is happy, and goes on trips with us. His identical twin brother is in New York with no access to this treatment is heavily medicated, also had life-threatening side effects from his medication, and due to his life-threatening aggression and self-injury, he has been unable to attend some family functions.

Does the UN care how individuals with disabilities are being treated in much of the world, where they have no rights to an education or any humane care at all?

In 1995, when I was in Thailand, I saw firsthand how individuals with special needs were being treated. I saw individuals with cerebral palsy with untreated contractures and shaved heads in a room with no beds, having their diapers changed on the same floor they were being fed. I witnessed individuals with other disabilities being treated exactly the same. Has the UN ever checked to see if the situation improved?

A year ago, when I was lecturing in China on autism diagnosis and treatment, I learned how individuals with severe intellectual disabilities and autism have no rights to any education. Unless their parents live in a city where there is a school and they have the funds to pay for it, the children either stay home or become institutionalized. Even the public schools for children with mild disabilities charge the parents money. Does the UN care that poor children with disabilities do not have access to humane care?

Today, the UN does not confront powerful regimes that murder their own people but instead targets Israel, as the Human Rights Council has condemned Israel in twenty of twenty-five resolutions. In the same way, rather than confronting powerful regimes that do not provide humane care to the disabled, to justify their existence, the UN calls a validated treatment being administered to individuals from not-so-powerful families "torture," as if these families who finally found help for their loved ones have not had enough stress in their lives. The UN wants to take away families' rights to choose effective life saving treatments for their children. Shame on the UN!

### *UN Convention on the Rights of Persons with Disabilities*

I was delighted when the UN Convention on the Rights of Persons with Disabilities was not ratified. I looked up the countries that did sign the convention. They included China, Thailand, and Afghanistan (where typical girls could not go to school). I wondered what this convention really meant.
The UN has condemned involuntary treatment even though involuntary hospital admissions save lives. "Much hinges on the [UN] Committee on the Rights of Persons with Disabilities' view that all persons have legal capacity at all times irrespective of mental status, and hence involuntary admission and treatment, substitute decision-making, and diversion from the criminal justice system are deemed indefensible."[290] According to this view, depressed individuals have the right to commit suicide, or people who are convinced their food is poisoned have the right to starve to death. In the absence of psychiatric illness, these people would not engage in such behaviors.

According to the UN Convention on the Rights of Persons with Disabilities, everyone had a right to an education, but obviously it was not enforced, and therefore, it was meaningless, except for a good political appearance. Yet the

290 Melvyn Colin Freeman, Kavitha Kolappa, Jose Miguel Caldas de Almeida, Arthur Kleinman, Nino Makhashvili, Sifiso Phakathi, Benedetto Saraceno, et al., "Reversing Hard Won Victories in the Name of Human Rights: a Critique of the General Comment on Article 12 of the UN Convention on the Rights of Persons with Disabilities," *Lancet Psychiatry* 2, no. 9 (September 2015): 844, doi:10.1016/S2215-0366(15)00218-7.

UN chooses to go after the JRC without ever visiting the school or speaking with a family member.

Of course, that is to be expected from an organization that in May 2013 placed Iran in charge of the UN arms control forum:

> Iran will preside over the United Nations arms control forum this month, despite the fact that it is under U.N. sanctions for illicit nuclear activities and routinely supplies arms to the terrorist organization Hezbollah in violation of international law. The U.N.'s annual Conference on Disarmament, which Iran is slated to lead from May 27 to June 23, is the organization's primary multilateral forum for negotiating arms control agreements.[291]

Furthermore, in 2015, Iran won a top post on the UN-Women's Rights Board, although women in Iran are arrested for not wearing head scarves correctly and cannot travel or receive an education without their husband's permission, among other human rights abuses.*292* Also in 2015, at a two-week UN meeting, Israel was the only country condemned for human rights violations against women.*293* Given the above, I think the UN position as an arbitrator of human rights is a joke.

I consider these anti-aversive groups to be dictatorial organizations because they want to dictate people's care according to their generalized standard, whether it is inclusion (even if the individual is just staring out the window) or using positive-behavior supports exclusively (even when it is ineffective, and the individual is a zombie from medication). They do not care about what the parent thinks. They believe they know better. However, certain

---

291 Alana Goodman, "Fox in the Hen House, Iran to Chair U.N. Arms Control Forum," *Washington Free Beacon,* May 13, 2013, http://freebeacon.com/national-security/fox-in-the-hen-house/.

292 "Iran Wins Top Seat on UN Women Board; Unlike Before, US & EU Fail to Back Alternative Candidate, Possibly in Deference to Nuclear Talks," *UN Watch,* April 13, 2015, 531, http://www.unwatch.org/eu-us-allowed-iran-to-win-top-seat-on-un-womens-rights-board-rights-group-says/.

293 "Israel Singled Out at UN for Women's Right Violations," *Times of Israel,* March, 21, 2015, http://www.timesofisrael.com/israel-singled-out-at-un-for-womens-right-violations/.

types of neglect and abuse appear to be acceptable to the government and the self-appointed anti-aversive advocates as long as they fit with, or at least do not go against, their purported agendas.

Not one of these "advocacy" groups have spoken out for the rights of those over the age of twenty-one to have access to due process to ensure appropriate care. Police sometimes use Taser or aversive sprays when they encounter a disabled person they believe may be a danger. When they do use these painful and sometimes injurious techniques, there is no professional specifying the indications, there is no internist or psychiatrist review, there is no legal guardian consent, no human rights approval, or court proceeding, unlike when JRC uses aversives. These same "advocates" do not advocate making these methods illegal if the police use them, even though police have killed with Tasers.[294] These supposed advocacy groups truly exclude the JRC families, who are some of the best advocates.

## The Food and Drug Administration and Aversives

In January 2013, the Food and Drug Administration (FDA) threatened to confiscate shock devices. When I saw the e-mail from the JRC, it just added another stress and more anxiety, as if I did not have enough, already. It never ends.

However, I immediately wrote a letter to the FDA. I thought of the medications the FDA has approved and their dangers, indicated by the lengthy over-the-counter drug package inserts, much less those for psychotropic medications. I thought about how for years they have allowed animals to be fed antibiotics on an everyday basis despite resistant infections. I thought about how physicians made patients suffer, including Talia, who once had pneumonia for a week until the diagnosis was confirmed, because doctors followed treatment guidelines on avoiding prescribing antibiotics because of resistant infections.

In April 2014, the FDA convened a hearing to ban the skin-shock device. I testified at this hearing. I informed them that Matthew went from two hundred

294 "Fatal LI Cop Tase," *New York Post*, July 26, 2013, http://www.nypost.com/p/news/local/fatal_li_cop_tase_Zmt4xThD4BPnrX9celD2LK.

applications in his first year at the JRC to none for the past two years and never had a side effect. I explained he was treated with a skin-shock device that was modified to be more intense after the initial one was not sufficiently effective.

The FDA initially told the school in 2000 that it did not need a clearance for the modified device, because it was covered under a practice of medicine exemption. However, the FDA reversed its position in 2011. The FDA presented as part of its justification three reported cases of former students with side effects such as posttraumatic stress disorder and seizures. The issue is that these cases were only heard by phone interview and without obtaining any information from a treating physician or therapist or with medical documentation. The device legality should not be subject to political whims.

One of the presenters at the hearing against aversives was Dr. Fredda Brown, who was, and continues to be, on the advisory board of Quality Services for the Autism Community (QSAC),[295] a day school. Two months after this hearing, QSAC rejected Talia because of her behaviors. Dr. Fredda Brown was the psychologist who designed the failed nonaversive program for the individual who died from medical consequences of self-abusive scratching (mentioned earlier). Dr. Brown also sat on the board of directors at TASH (an anti-aversive organization described elsewhere).[296] She has advocated against the use of aversives in Albany as well. She contributed to the Mental Disability Rights International (MDRI) report,[297] a report so desperate for information that it took bits and pieces of a testimonial of a former student from the JRC website to give the impression he was abused. I have met this student around the country as he advocated for the right to aversives with me, including testifying at the FDA panel that skin shocks saved his life.

At the hearing, Dr. Wayne Goodman, on the FDA advisory panel, asked if placebo-controlled trials on the skin-shock device were done. When later

295 "Board of Directors," Quality Services for the Autism Community, accessed February 7, 2016, http://www.qsac.com/who-we-are/board-directors/.

296 "Fredda Brown," TASH, accessed February 7, 2016, http://conference.tash.org/speakers/fredda-brown/.

297 Laurie Ahern and Eric Rosenthal, "Torture, Not Treatment: Electric Shock and Long-Term Restraint in the United States on Children and Adults with Disabilities at the Judge Rotenberg Center," *Mental Disability Rights International*, 2010.

interviewed for an article, "[He] stated that though many patients have seen benefits from this treatment, he is 'not ready to say we should disseminate this treatment to others.'"[298]

First, it may be unethical to give a placebo to someone with acute life-threatening behavior. Second, an observer could not really be blind to who was given a skin-shock application, when the person's reaction would be observable. I made a vocalization when I tried the shock.

Furthermore, placebo-controlled trials, while important, have limitations. Even if results are statistically significant, they may require treating a large number of people before finding a clinically meaningful response. In addition, if the sample size is large enough, there may be a statistically significant result, even if the magnitude of change is small. That is not acceptable for a life-threatening behavior. Single- and multiple-case studies with baselines reporting actual frequencies of a given behavior have utility to demonstrate individual clinically meaningful change and are often used in behavior analytic research, including regarding skin-shock treatments.

An example of a useful medication that never required a placebo-controlled trial is penicillin. In 1942, penicillin was experimented on a woman who was near death after a blood infection subsequent to a miscarriage.[299] She was cured. If she was placed on a placebo, she would have died, and a placebo was never necessary to demonstrate the effectiveness of this life-saving medication.

Although some individuals with cognitive impairments such as my brother do not have the capacity to provide an informed consent for treatment with skin shock, these individuals, just like children, do not have the capacity to consent for any medical procedure. However, it is unethical to withhold treatment because of a disability that interferes with judgment. Rather, as with children, the legal guardians need to provide informed consent.

---

298 "FDA Panel Considers Ban on Electric Stimulation Devices for Aversion Therapy," *Healio Psychiatric Annals*, April 24, 2014, http://www.healio.com/psychiatry/practice-management/news/online/%7B35d99d00-ff2e-46c2-9368-2e11ded4fa39%7D/fda-panel-considers-ban-on-electric-stimulation-devices-for-aversion-therapy.

299 Howard Markel, "The Real Story Behind Penicillin," *PBS Newshour*, September 27, 2013, http://www.pbs.org/newshour/rundown/the-real-story-behind-the-worlds-first-antibiotic/.

On April 22, 2016, the FDA announced a proposal to ban the skin-shock treatment entirely for self-injury and aggression.[300] In the press release, Dr. William Maisel, acting director of the Office of Device Evaluation at the FDA's Center for Devices and Radiological Health, since 2014, stated, "Our primary concern is the safety and well-being of the individuals who are exposed to these devices...These devices are dangerous and a risk to public health-and we believe they should not be used." The press release also stated, "The FDA believes that state-of-the-art behavioral treatments, such as positive behavioral support, and medications can enable healthcare providers to find alternative approaches for curbing self-injurious or aggressive behaviors in their patients," despite evidence to the contrary, and that in 2014 the FDA advisory panel unanimously agreed that no other option existed for these behaviors. Furthermore, the ban is limited to use for treating self-injurious or aggressive behavior, as stated in the news release. It does not include using electrical stimulation devices as an aversive to treat addictions. Individuals who want to quit smoking, stop their binge eating, or other behaviors can purchase Pavlok, an aversive electrical stimulation device online without any prescription, and according to the website, "Pavlok vibrates, beeps, and releases an electric stimulus that ranges from pleasant to aversive."[301]

In 2012, Dr. William Maisel was arrested and pled guilty to being part of a ten-men prostitution ring after offering an undercover police officer money for sex.[302] He only had to pay a $200 fine,[303] which is less than some parking tickets. Apparently, his probation consisted of his new position at the FDA. In

300 "FDA Proposes Ban on Electrical Stimulation Devices Intended to Treat Self-injurious or Aggressive Behavior," *FDA News Release*, April 22, 2016, http://www.fda.gov/NewsEvents/Newsroom/PressAnnouncements/ucm497194.htm.

301 "Pavlovian Conditioning on Your Wrist," *Pavlok*, accessed April 22, 2016, https://pavlok.com/.

302 Larry Husten, "Cardiologist William Maisel Arrested in Prostitution Sting Operation," *Forbes*, August 1, 2012, http://www.forbes.com/sites/larryhusten/2012/08/01/cardiologist-william-maisel-arrested-in-prostitution-sting-operation/#3b3c5521397f; Nalini Rajamannan, "FDA Official, William Maisel, Pleads Guilty to Crime, Keeps Job," *Laurel Patch*, October 29, 2013, http://patch.com/maryland/laurel/fda-official-william-maisel-pleads-guilty-to-crime-keeps-job.

303 Rajamannan, "FDA Official, William Maisel, Pleads Guilty to Crime, Keeps Job," *Laurel Patch*, October 29, 2013.

my opinion, someone with a criminal record should not have the authority to determine the legality of my brother's or anyone else's treatment for that matter.

According to one article, Dr. William Maisel was also involved "in a scandal in which FDA officials spied on the agency's own scientists who had expressed concerns about the safety of medical devices."[304] The article further states that Dr. Maisel reportedly "was responsible for the final decision to fire Ewa Czerska, one of the targets of a surveillance operation in which the agency monitored the computers and communications of a group of scientists who were subsequently fired or left the agency and later sued." It appears that although Dr. Maisel is opposed to aversives, he along with other FDA officials, have caused problems for scientists at the FDA who had safety concerns of other medical devices.

It is also no surprise that the FDA press release suggests medications. The current FDA commissioner, Dr. Robert M. Califf, "has deeper ties to the pharmaceutical industry than any F.D.A. Commissioner in recent memory," and a "financial disclosure form…listed seven drug companies and a device maker that paid him for consulting and six others that partly supported his university salary, including Merck, Novartis and Eli Lilly. A conflict-of-interest section at the end of an article he wrote in the European Heart Journal…declared financial support from more than 20 companies."[305]

According to one report, he led a clinical trial of Xarelto (ROCKET AF) (rivaroxaban), in which "a senior FDA official wrote that a 'lack of care' in ROCKET AF's design and execution might have led to avoidable strokes among test subjects," and that the trial "relied on a blood-testing device with a history of malfunctioning and delivering false results."[306] In fact, according to a different article, "[j]ust as the trial was getting underway in 2006, the

---

304 Emily Heil, "FDA Official Caught in Prostitution Sting," *Washington Post*, August 6, 2012, https://www.washingtonpost.com/blogs/in-the-loop/post/fda-official-charged-in-connection-with-prostitution-sting-watchdog-raises-questions-about-his-role-in-spying-scandal/2012/08/06/8e943486-dfd7-11e1-a421-8bf0f0e5aa11_blog.html.

305 Sabrina Tavernise, "F.D.A. Nominee Califf's Ties to Drug Makers to Drug Makers Worry Some," *New York Times*, September 19, 2015, http://www.nytimes.com/2015/09/20/health/fda-nominee-califfs-ties-to-drug-industry-raise-questions.html?_r=0.

306 "New FDA Head Led Flawed Clinical Trial," *POGO: Project on Government Oversight* 20, no.1 (January-March 2016): 1.

INRatio [blood testing device] was facing scrutiny by the F.D.A. In 2005 and 2006, the agency sent warning letters to HemoSense, then the manufacturer of INRatio, claiming that the devices were generating 'clinically significant' erroneous values and that the company, which was later acquired by Alere, was not properly investigating the complaints."[307] The article further states that Alere recalled the device in 2014. Another article stated, "Since the INRatio and a later model, the INRatio2, were cleared for use in 2002, the F.D.A. has received more than 9,000 reports of malfunctions with the products, and more than 1,400 reports of injuries, according to an analysis in December by the Public Citizen Health Research Group, a consumer organization."[308]

In summary, while the FDA has expressed concern with an aversive skin-shock device, the current FDA commissioner led a clinical trial using a faulty device despite reports of injury.

The alternative to aversives always leans on medications. Medications have side effects and cause death. Medications were never helpful to Matthew. Even the FDA panel, with the majority supporting a ban on skin shock, unanimously agreed that no other option existed for individuals who had behaviors like Matthew's. The shock device should not be held to a different standard than medication, especially when the risk of no treatment can be death and off-label polypharmacy adverse effect trials have never been done for this population. Some individuals with autism could not even receive a physical exam until the aversive shocks were used to manage their behavior.

## Poetic Advocacy

I get furious at judgmental parents who think, because their child did not need a particular intervention, that no child will need it, as if we no longer

---

307 Katie Thomas, "F.D.A. Asks if Faulty Blood Monitor Tainted Xarelto Approval," *New York Times*, February 22, 2016, http://www.nytimes.com/2016/02/23/business/fda-asks-if-faulty-blood-monitor-tainted-xarelto-approval.html.

308 Katie Thomas, "Accuracy Concerns on Testing Device for Blood-Thinning Drug," *New York Times*, March 17, 2016, http://www.nytimes.com/2016/03/18/business/accuracy-concerns-on-testing-device-for-blood-thinning-drug.html.

need to individualize treatment. My mother and I wrote a poem on the blog of one parent who is an anti-JRC, anti-aversive activist, and a professor.

"No Life"
You are fortunate that your child
Only required interventions that were mild.
JRC's therapy saved my son's life.
No longer would his self-injury require a surgeon's knife.
Without the therapy, I will stay up at night
With no end to my stress in sight.
While you sleep well in your bed
As my son rams his head
And then drops dead.
Ignore 112 peer-reviewed articles on skin-shock at my son's peril. You know better!

## Advocacy in State Houses and Congress

When I was twenty years old and on a trip to visit Matthew, my mother was asked to advocate against pending Massachusetts legislation to stop the use of aversives the day after we arrived. She was unable, so I volunteered. This was my first experience with advocacy. At first, the director of Matthew's school, Dr. Israel, declined my offer, but later that day, he changed his mind. He was not sure if I would be effective. I testified at a public hearing at the State House Human Rights Committee. One of the representatives later told the school staff that because of my testimony, he would never vote to stop the use of aversives. This gave me confidence to continue.

Now more than twenty years later, I have spoken to countless congressional aides, state assemblymen, and senators and their aides in New York and Massachusetts. Once I had to go to Albany, New York, for a day while I was visiting Matthew in Massachusetts. With Talia's autism, I do not travel much anymore, so I visit Matthew only once a year. It was frustrating to have to spend one of those precious days in Albany.

On another occasion, I had the most stressful trip on a visit to Washington, DC. I went there because a proposed bill would ban aversives and allow only restraint and seclusion as part of a general policy for dangerous behavior. This was part of an effort by the Alliance to Prevent Restraint, Aversive Interventions and Seclusion (APRAIS).

I had to catch a plane on a Tuesday night after leaving work. That day, my doctor called to inform me that a biopsy result was inconclusive, so I might have breast cancer. Between patients, I was trying to make an appointment with a breast surgeon, as my doctor instructed. I was running behind with my patients and even had to reschedule one or two of them. I ended up working until about six o'clock in the evening and dashed down to the airport. I accidentally went to the wrong terminal, which was only for my return flight, and then had to get to the correct terminal. I arrived in Washington, DC, at about ten thirty at night, exhausted and stressed out. I was wondering if I had cancer; if I did, I wondered who would take care of my children, what would happen to them if I died, and if Talia would be able to stay at her school.

I spent the next two days speaking with senators' aides, repeating my story about every thirty minutes. Some appeared sympathetic; in particular I liked the aide in Senator Lisa Murkowski's (R-Alaska) office. She said that with a ban on restraints and seclusion on IEPs, parents would have no say in the procedure if their child needed such an intervention and that the only say a parent had in their child's education was through the IEP. Some students have physical problems, and it is only safe to restrain them in certain ways, and that could not be written on an IEP with this legislation.

I remember after speaking with her, I went into the hallway and called my house, trying to participate in a home team meeting regarding Talia. As I visited other legislators, I found Senator Tom Harkin's (D-Iowa) office unsympathetic; he has since sponsored equivalent bills. Senator Christopher Dodd's (D-Connecticut) office was particularly cold and uncaring; he was a sponsor of the bill.

His was our last meeting. I walked out with another parent, and we discussed how we were both emotionally drained. I came home late at night, and the next morning, I met with my breast surgeon. I picked a date for a lumpectomy and reconstruction. No one should have to go through all this.

In December 2011, the OPWDD wrote proposed regulations to prohibit the use of aversives. A group of parents, school staff from the JRC, and I left our homes at five o'clock in the morning to meet with staff from the governor's office and later with some legislators and OPWDD staff. That morning, Governor Cuomo's office canceled our meeting, stating they would meet only with the lobbyist, not the families. We did meet with some legislators. I surprisingly found the OPWDD commissioner, Courtney Burke, to be receptive to us.

One of the most difficult months was June 2012. Anderson Cooper of CNN News gave some negative reports on the JRC after a parent sued for alleged mistreatment in 2002. Cooper clearly did not do his research beforehand and even made the comment that most of the research on skin shock was done at the school. This was not true, and the JRC's research director corrected him.

On one of his shows, broadcast May 18, 2012, Cooper even had a JRC former employee berate the JRC, but he never disclosed that this person had been demoted, refused the new position, and wrote a letter asking for his former position back. While this individual worked for the JRC, he evaluated other staffers and wrote that they were not administering the skin-shock device *enough* in accordance with clinically approved protocols. This individual was also interviewed for CBS in the 2014 report mentioned earlier, and this important information about him was never disclosed.

After watching this negative reporting, I had a nightmare:

> I was in the OPWDD commissioner's office trying to stop regulation that would ban Matthew's treatment, but the commissioner was not there. I saw on a desk the textbook, *Effective Programs for Treating Autism Spectrum Disorder: Applied Behavior Analysis Models*-the textbook Matthew's school was listed in as a model program. It had a big black "X" on the cover. I left with some parents and staff and went into the hall to some pay phones, trying to call anyone who would speak to us, anyone who would help.

A few weeks later, I was back in Albany to advocate against a bill Governor Cuomo wanted passed and which Autism Speaks had promoted. I could not think of a more stressful trip. Bill S7400 was the bill to create a "Justice Center" with a mission "supporting and protecting the health, safety, and dignity of all people with special needs and disabilities through advocacy of their civil rights, prevention of mistreatment, and investigation of all allegations of abuse and neglect so that appropriate actions are taken."[309] The Justice Center was a response to allegations of mistreatment and abuse in some articles in the *New York Times*.

The governor appointed a special advisor, Clarence Sundram, who had a long history of opposition to aversives. He was involved in the Mental Disability Rights International (now known as Disability Rights International, DRI) report discussed earlier. I believe it distorted the JRC and its treatments, even taking words out of context to give an appearance of abuse from a former student who stated that skin shocks saved his life.

In an interview in Albany on *Your News Now Capital Tonight*, on June 20, 2012, Sundram made the false statement that "every professional organization" (and he named some) "has adapted resolutions against aversives."[310] While some have adopted such resolutions, other professional organizations have not, such as the Association for Behavior Analysis International. The JRC is an organizational member of that association, and staffers from the JRC have spoken at their annual conferences on aversives.

Sundram went on to state, "There are lots of other interventions which you can use to prevent people from harming themselves."[311] While positive interventions can be effective for some, the research clearly shows their limitations. To further this problem, the programs that exist in New York use underfunded and therefore poorly implemented positive-behavioral interventions. That is why so much off-label prescribing of potentially toxic medication occurs.

The bill was eighty pages long, it would create a "Justice Center" on abuse, and it contained a few lines equating aversives with abuse. It would deny funding to any program that used aversives, unless the funding agency-in this

309 New York State Justice Center for the Protection of People with Special Needs Mission Statement, accessed February 14, 2016, http://www.justicecenter.ny.gov/about/vision.

310 Clarence Sundram, interview with Liz Benjamin, *Your News Now Capital Tonight*, New York, June 20, 2012.

311 Ibid.

case the OPWDD-agreed to the use of aversives, which would most likely never happen. I think the bill was only lip service, because it simply created another government bureaucracy to investigate abuse rather than an independent body, and it would do nothing to remedy the preexisting risk factors for abuse, such as shortages of workers, poorly trained and underpaid staff, and limited supervision.

Five days before Sundram's interview, a few JRC families, a former student, staff from the JRC, and I spoke with representatives from the assembly and senate. The answer we received was that this bill was from the governor's office, and no one from the assembly or senate would speak up to the governor, even when we explained that no one in the governor's office spoke to us.

No one from the governor's office would speak with us, so the governor got to hear us speak about him. On June 15, I did radio and television interviews with another parent, and the JRC's research director also appeared on the television show. Our lobbyist was almost sure the governor would be watching the television show.

Going on television, I experienced palpitations; I would be begging for my brother's life. Right before going on air, I tried to picture flowers in my mind to calm myself down. I thought this may be my last chance to get my point across.

I explained that the therapy saved Matthew's life. While the adverse effects of psychiatric medications have killed people, it would be irresponsible to deny these medications to everyone, because the medications also save lives and allow some people to hold jobs and care for their families. In the same way, it would be irresponsible to ban aversives for everyone, even though there have been problems with them in the past.

On the television show, I felt so hopeless. At home that night, I had nightmares of loss of control:

> I was hanging to the edge from a wing of a plane way above the ground. I was losing my grip. I thought about grabbing my older daughter, Batsheva, who was sitting more securely on the plane, but I did not, worrying I would endanger her life by doing so.

I also dreamed I was with Talia in the street, and a tornado was coming with no place for us to take cover.

Going into work after this trip, I saw my patients, with their side effects to their medications, and how they still engaged in dangerous behaviors. I thought, "This will be Matthew." I thought how, despite all my training, I could not protect him. I started to think I could no longer work with this population-it was too upsetting-and that I could not give them what they really needed.

The senate passed the governor's bill in three days. I doubt anyone ever read the eighty-page bill prior to that vote. I was informed that the senate, with its narrow Republican majority, wanted to please the Democratic governor, so he would not campaign against them and risk their majority. The governor announced that the bill had to be passed before the legislators could go on vacation.

However, in the end-and surprisingly-two assemblypeople stood up to the governor: Amy Paulin and Felix Ortiz. On the assembly floor, on the record, Assemblyman Ortiz answered Assemblywoman Paulin's questions, which made it clear individuals receiving aversives at JRC would be able to continue to do so. I felt so relieved. After the bill was passed, a joint statement by Sundram and Bob Wright (from Autism Speaks) praised the Justice Center.[312]

However, the Justice Center legislation, as I suspected, did not address the real needs. A Justice Center hotline is of no use when so many of the disabled have limited communication skills and cannot testify to their abuse. If I tell someone who never saw a lion that lions do not have manes, then lions do not have manes. If the abuse is not seen or clearly communicated, then it did not exist.

More than a year later, this article on the Justice Center issued the following statements:

> A review by *The Times* found that the state had made no discernible progress in firing abusive and derelict workers...[ Governor Cuomo's]

312 Clarence Sundram and Bob Wright, "And Justice for All in N.Y.," *Times Union,* June 24, 2012,
http://www.timesunion.com/opinion/article/And-justice-for-all-in-N-Y-3658415.php.

> appointee to lead it [the Justice Center], Jeffrey Wise, has alarmed some advocates for disabled people: Mr. Wise is a longtime spokesman and lobbyist for private disabled-care providers, who are often as troubled as the state…Mr. Wise even lobbied against Jonathan's Law, the legislation that forced the state to start disclosing abuse reports to parents, named after a teenager with autism who died after being asphyxiated by a state worker.[313]

In another article, despite the presence of the Justice Center, OPWDD allegedly did not respond for weeks after being notified of physical abuse occurring at one of its own residences.[314] Reportedly, individuals were hit, kicked, spat on, denied food, and lacked effective medical care.

In 2014, a few other families, including ours, tried to get language included in the New York budget bill to give individuals over twenty-one who were in out-of-state placements the right to a hearing before transferring them to in-state facilities (the same rights as in-state people already had if there was a proposed transfer of placement). Governor Cuomo would not allow it and had the assembly and senate draft wording removed from the bill. We contacted Autism Speaks beforehand regarding this legislation, and the organization chose not to get involved.

However, we continued to advocate. On November 21, 2014, the governor signed legislation allowing for due process procedures to contest a proposed placement to extend to individuals currently residing in out-of-state facilities.[315] Although the "due process hearing" consists of a panel of OPWDD employees by OPWDD, there is a right to appeal to a federal judge.

In conclusion, for some individuals applied behavior analysis approaches including positive behavioral intervention and supplementary aversives, if

---

313 Danny Hakim, "New York State Lags on Firing Workers Who Abuse Disabled Patients," *New York Times*, August 8, 2013, http://www.nytimes.com/2013/08/09/nyregion/state-lagging-on-dismissals-in-abuse-cases.html?_r=0.

314 "Lawsuit: Abusive Staff Called Group Home The 'Bronx Zoo,' Treated Residents Worse Than Animals," *CBS New York*, May 2, 2016, http://newyork.cbslocal.com/2016/05/02/bronx-group-home-abuse-lawsuit/.

315 New York State Mental Hygiene Law, section 13.37: 13.38 (2015).

needed, are safer and more effective than medication management. Unlike a two-second skin shock, medications remain in the body. Applied behavior analysis approaches also improve functioning and quality of life. There needs to be adequate staffing, training, and supervision to implement intensive behavioral approaches. Video monitoring ensures treatment fidelity while preventing neglect and abuse. While there are individuals who need psychiatric medication, we must end the practice of medication substituting education.

# Six

## Recommendations

In this chapter, I discuss recommendations to obtain effective treatment for individuals with developmental disabilities, to improve their symptoms, functioning, and quality of life. General recommendations and specific recommendations to families are explored.

### General Recommendations

Preventing neglect and abuse requires sufficient funding for adequate staffing, staff training, supervision, and monitoring. Staffers are often not trained properly.

In addition, at any facility where individuals cannot testify in a court, cameras are a necessity for their safety. Twenty-four hours a day and seven days a week video monitoring and the rights to unannounced access by legal guardians have always been standard at the JRC and should be at every school or placement for older individuals. Just as cameras have been effective in stopping people from going through red lights, people would be more likely to think before they acted if they were being filmed. Staff members would be unlikely to spend much of their day on cell phones, texting, or reading a newspaper if they knew they were being filmed. Cameras would also prevent residents from harming each other while staff were busy with other residents. Cameras could also reduce the need for 1:1 staffing for residents with self-injurious behaviors, sexually inappropriate behaviors, or escape risks, which would save money and be less restrictive to individuals.

Cameras would also protect staff from false accusations from higher-functioning residents (and in one instance I am aware of, another coworker). In some circumstances, the staffer is put on paid leave until the accusation is investigated and, normally, found to be groundless. Tax money is wasted because there are no cameras.

However, per diem staffers are often put on unpaid leave. One staffer in particular was accused of abuse by an individual. The accusation was later found to be unsubstantiated. The staffer told me he could have lost his apartment.

Of note, I would like to ask any of the politicians or self-appointed advocates so quick to condemn aversives (although most never visited the JRC) and who profess their concern about individuals with special needs, if they ever visited a local special-education program, day-habilitation program, or residential program unannounced. Do they ever have any concern about poorly done functional behavior assessments (when one is even done), inconsistent behavior plans at programs, or that programs coerce parents to medicate their children with potentially toxic drugs? I have never heard of one politician or self-appointed advocate who did.

Training and supervision are imperative not only to prevent neglect and abuse but also to carry out consistent therapies in the way they were designed and have shown evidence of success in research. Otherwise, these therapies can be used in modified, untested versions, making ABA and TEACCH ineffective.

Staffers need training even on how to speak properly to patients. I have known one direct-care worker for a long time, and I believe he cares about the people in his care. He once teased one of my patients, saying, "The doctor is going to give you more medicine," which only agitated the individual. I have asked the worker not to speak like that, but he should have been taught in his training not to tease.

That same week, I saw a well-meaning staff member from a residential-school setting reassure a fearful patient that the staffer would not document abuse she was required to report. I later explained that although her intention was good, in that she meant to calm the patient down, the patient also needed to establish trust and would not be able to do so if she lied to the patient about the protocols.

To prevent injuries during restraint, training should be mandated for any classroom staff with a student with a behavior plan on an IEP. This training should use a formalized crisis management procedure, such as "Physical/Psychological Management Training," which "looks at a crisis episode through four stages and helps to increase staff's understanding about their responses to the crisis, the interventions options available to them, and the management of their individual fear during a crisis situation…Included in this training are prevention and planning strategies; communication techniques of re-direction, verbal cues, and DE-escalation strategies; the quick-action teamwork necessary to maintain a safe environment and the post-event debriefing that allows for all staff to learn from the situation. Also included are the more restrictive techniques involved in escorts, seclusion, and physical restraint."[316]

It should be illegal to keep developmentally disabled individuals in windowless rooms. Similarly, facilities should not sell them junk food that impairs their physical and mental health. Artificial food dyes need to be banned for the same reasons.

If a physician does not inform a parent that a child needs antibiotics for an infection, that doctor could be professionally liable for malpractice. However, there is no personal liability for Committee on Special Education members if they do not provide for parent training, despite the research or regulation, or other needed treatments in the IEP. The IEP committee members make decisions that can have a significant impact on a child's life. They should have the same malpractice liability that a physician has, to make them accountable for their recommendations.

316 "About Physical/Psychological Management Training PMT)," Capitol Region Education Council, accessed February 14, 2016, http://www.crec.org/ppt/.

Schools also have no liability when a child regresses because they did not implement evidence-based treatment. Again, they should have the same malpractice liability that a physician has, to make them accountable for providing an appropriate education.

Curriculums need to be individualized for each child. Following a state's core curriculum is harmful, if a child does not have the cognitive capacity to understand and learn from it. Imagine if you had to be seated all day listening to a teacher speaking in Alaskan Eyak. Imagine that you could not speak, that you could not say you have no idea what that teacher is talking about. Five days a week, for hours every day, you must sit there and listen to it. Forcing disabled children into programs that they cannot understand only serves to increase their frustration, self-stimulatory behaviors, aggression, and self-injury-and consequently, their use of dangerous psychotropic medications.

Instead, a curriculum should be a hierarchy of obtainable and measurable goals and objectives based on a child's level of functioning, not blindly cutting and pasting from a template, which has been done. Examples of a measurable goal are as follows:

- Will be able to put toothpaste on toothbrush with one verbal prompt.
- Will expressively identify numbers one through five.
- Will receptively identify three new words in a field of five with one verbal prompt within three seconds.

Schools need to focus on total functioning rather than making test scores the only important thing. We need to raise functional adults who can stand on their own two feet and who can take care of themselves and work. Just doing well on exams does not cut it.

Activities in day programs and residences for people over the age of twenty-one need to be cognitively and developmentally appropriate rather than age appropriate. "Age appropriate" is a generalized concept to apply to anyone within a certain age range. "Cognitively and developmentally appropriate" means activities are chosen on the basis of the specific individual's cognitive level and developmental stage. Furthermore, "age appropriate" is subjective, up to an individual's judgment, and cannot be measured-like an IQ score, for example.

At a day-habilitation program where I was previously employed, my supervisor informed me that OPWDD had plans to phase out day-habilitation programs in classroom settings. She also informed me that OPWDD was encouraging what is called day habilitation without walls, where individuals are in the community or at home during the day accompanied by staff. It does and will save money on building space. However, going into the community daily or staying at home all day can decrease a person's structure and routines, and these help with autism and other disabilities. If an individual has a communication impairment, rather than going to places of interest to that individual, the staffer may go wherever the staffer wants. Again, these programs need to be appropriate.

In 2015, OPWDD encouraged individuals living in group homes to have their own apartments. While I have witnessed some successes, others have been incapable of attending to their own medical needs, such as taking their medications consistently resulting in significant and even life-threatening regression. Residences also need to be appropriate to the specific needs of the individual.

People should not lose their rights to effective, appropriate treatment just because they reach the age of twenty-one. All the hard work done by parents and professionals should not be thrown out the window by stopping treatment at the age of twenty-one, causing the person to regress. The government should not be allowed to stick someone with a disability in whatever place it pleases and call it appropriate or provide no vital services at all. We need rights to impartial hearings for people over the age of twenty-one.

Criminals have rights to an attorney, but disabled children do not. The JRC gives families free legal assistance, and the DOE unsuccessfully tried in federal court to stop it.[317] Parents are not professionals and cannot be expected

317 AR, as parent and natural guardian of NB, a minor, plaintiff, v. New York City Department of Education, US District Court, Southern District, no. 12, Civ. 7144, October 28, 2014, WL 5462465, http://www.leagle.com/decision/In%20FDCO%2020141029E36/A.R.%20v.%20NEW%20YORK%20CITY%20DEPARTMENT%20OF%20EDUCATION.

to be familiar with all the regulations. School districts have attorneys. Giving children rights to legal representation would help solve the disparity and segregation in the quality of education between what children of poor and rich families receive. It would encourage the creation of appropriate public school programs to avoid litigation and would serve all children equally. It would help stop overmedication. These rights can also help older individuals.

For children or adults with developmental disabilities, we need to think of behavior in terms of a symptom of a medical condition, just as difficulty breathing is a symptom of asthma. For example, if a parent brought a child with asthma to a hospital because of difficulty breathing, it would be unacceptable and even malpractice if a doctor told the parent that he or she could not handle the problem and to take the child home. In the same way, it is unacceptable and malpractice if someone with a neurologic condition has a dangerous behavior and the school or adult program tells the parent to take the child home.

However, unlike with asthma, no medication improves functional skills for autism. It is the educators and the behavior analysts who are the true doctors for autism and other conditions. "ABA is a primary method of treating aberrant behavior in individuals who have autism. The only interventions that have been shown to produce comprehensive, lasting results in autism have been based on the principles of ABA."[318] If the staff cannot manage the behavior, they need to refer the family to a placement that can, and not just dump the problem onto the parent.

We need ABA and crisis day programs for individuals with developmental disabilities when standard day programs cannot manage their behaviors. This would provide an increased level of structure and supervision to help stabilize behavior rather than simply dumping the problem onto a parent, who may be elderly. We also need residential programs in which staffers are trained to handle violent behaviors, so parents do not have to keep violent individuals at home, where these individuals may threaten their parents' lives.

---

318 Richard M. Foxx, "Applied Behavior Analysis Treatment of Autism: The State of the Art," *Child & Adolescent Psychiatry Clinics of North America* 17, no. 4 (October 2008): 821, doi:10.1016/j.chc.2008.06.007.

In the context of discussions of autism treatment, "human rights" often means relying on toxic medication. We must approach these situations holistically. A true human rights abuse is to place someone on medication-with potentially life-threatening side effects-to control the person's behavior rather than using a benign intervention such as withholding a CD player or access to money as a consequence to decrease the frequency of a dangerous behavior. Parents perform these types of benign interventions with their children.

As stated before, I withheld Stuart's train books after he banged on a bathroom door at our synagogue because he felt he had to use that particular bathroom. He never did that again. By contrast, at one program, a staffer was reprimanded for removing an item from a participant who was about to throw it at someone else, despite a prior verbal prompt not to do so. Such benign interventions, even to protect others, are considered a "human rights abuse." A simple intervention that may significantly help with behavior is making sure individuals get enough sleep, which may include enforcing turning off electronic devices after a certain time of day or night.

In the name of human rights, individuals in residences are allowed to refuse not only a day program but also doctor's visits. If a child refused a doctor's visit, we would not accept the refusal, because the child does not have good judgment. Yet individuals with disabilities often do not understand their medical conditions or need for treatment, so it is neglectful to allow them to refuse a doctor's appointment. It should not be any more of a "human rights abuse" to compel necessary medical evaluations and treatment for individuals with impaired judgment than giving children vaccines.

In medical school and physician residency training programs, trainees have to work with different supervisors in different settings. However, BCBAs are not required to work in more than one setting with more than one supervisor. The medical profession has specific published treatment guidelines for specific disorders. By contrast, while multiple textbooks and over a thousand peer-reviewed scientific published articles support ABA, specific treatment guidelines with algorithms do not exist for specific behaviors. A BCBA once told me she needed to terminate with a client and did not know exactly how to go about it. I showed her psychiatry patient termination guidelines. However, at least BCBAs have a minimal standard of required education and training and must take an exam. At some programs, the individual who supervises the ABA may not have a BCBA or otherwise have adequate credentials to provide that form of therapy.

It is up to the parents to screen and evaluate their children's therapists. In New York City, when a child turns three and switches from early intervention to the school district, unless the parent pays for private therapy and seeks reimbursement through an impartial hearing, it is difficult to terminate an incompetent ABA therapist. Sometimes, they have no one supervising them. When their children are above the age of five, parents in New York City are left completely on their own to find private therapists.

We must improve the BCBA standard, on the basis of the medical model's training and treatment guidelines. BCBAs should be mandatory at programs for anyone with dangerous behaviors, just as special-education teachers are required in classrooms for children with disabilities.

The behavior analysts who conduct the functional behavioral assessments and behavior plans should work at both the residence and the day program of a given individual to promote consistency. Even for Talia, school staff come to our home for parent training and participate in team meetings with home staff. Our home staff observe her at school and monitor videos of her at school as well. This way, everyone is working together for consistency.

Individuals with dangerous behaviors or behaviors that interfere with learning need the right to a functional experimental analysis, if other methods cannot determine the function of the behavior. If a school or adult program refers someone to a psychiatrist, it should also automatically consult with a board-certified behavior analyst to minimize chemical restraint. In addition, to prevent overmedication of people with developmental disabilities, we need a federal law to regulate it, just as there is one for nursing homes. Residential facilities must be required to keep summary records on previous medication trials, reasons for use, dosages, durations, and reasons for discontinuation to further optimize care.

Just as part of psychiatric training is performing psychotherapy, psychiatrists who treat individuals with developmental disabilities need training in applied behavior analysis and should visit successful applied behavior analysis programs. They need to see individuals who have had dangerous behaviors but learned through an effective, behavioral treatment program to be safe and productive citizens without the use of psychiatric medication. This would help these professionals advocate for appropriate educational programs, as opposed to off-label prescriptions of psychotropic medications. We should strive to start dual-certification programs in psychiatry and applied behavior analysis.

It is economical to provide free and appropriate education for individuals with autism or other disabilities, because in the long run, these individuals can be productive, which would save money. Currently, education funding is limited, but medications are funded much more liberally. If I prescribe 1:1 ABA or increased speech therapy to help with communication and behavior, the parent will likely have to go to a hearing and may need to pay for a lawyer, private evaluations, and the therapy itself-and later seek reimbursement. Whereas, when I write a prescription for medication to control behavior, Medicaid will never have the family overcome such barriers in order to have access to medication.

It would help if educational-behavioral techniques came out of the same budget as medication management. The toxic pill would not be prescribed so readily as a substitute for ABA or speech and language therapy to manage behavior. The most important thing for a person with developmental difficulties is to learn to behave. It does not matter how high functioning people are; if they cannot behave, they cannot function.

Staff members in schools and adult programs need training in infection control. Schools must make sure children's hands are washed before they eat, as is a required procedure in Japan. It is disconcerting to visit a placement and watch individuals eating meals without washing their hands, especially special-needs persons-who may put their hands into some very unsanitary places. Staff should also not be allowed to handle the food with bare hands. Furthermore, when individuals with disabilities have infections, they may express their discomfort through behaviors. Fewer infections mean better school and day-program attendance. In New York State, at least, attendance drives funding.

Professional testimony at hearings or other court procedures is not a luxury. People need these services. Whether the professional is privately paid or is employed by a Medicaid clinic, individuals' lives may depend on the testimony. It is unethical for professional testimony to be a boutique service.

Individualized Service Plans need legally mandated, appropriate staffing ratios for residences and day programs. If a facility is out of compliance, as sometimes happens because of budget cuts, or if there is reason that an individual needs a higher staffing ratio than on the ISP, the parent or other legal guardian must have the right to a hearing.

Just as schoolchildren and employees learn about racial and religious tolerance, the general public needs to be educated about individuals with disabilities. The words or actions of even one person could have a tremendous impact on others. If Stuart had not been teased at work, he might be a functioning, productive individual today, and my family would have been saved much sadness and grief.

Using private health insurance to fund autism treatments has many limitations, the greatest being parents' lack of private insurance or their inability to afford the copayments and deductibles. When parents lose their jobs-and their health care insurance-children lose access to their autism treatments. Advocacy groups would do better to push for a strengthened Individuals with Disabilities Education Improvement Act and to pressure school districts to do what is right, without forcing parents to pay for private school tuition, attorneys, and therapists out of pocket. Parents then have to sue for reimbursement. We should make education free and appropriate for all, to include home ABA and other autism therapies, when needed.

Everyone should have the right to live in the state of his or her choice. Matthew and Stuart, and their legal guardians, want both my brothers to live nearby in Massachusetts. However, Stuart is forced to live in New York, and New York State has tried to remove Matthew from Massachusetts. Massachusetts will not pick up the funding for the JRC if New York terminates it. If a person without a disability moved to any state and qualified for a state program such as Medicaid or welfare, the person would get it. Just because my brothers have certain developmental disabilities should not mean the government gets to dictate the state in which they have to live. People with disabilities or their legal guardians should have a choice over what state to reside in, just like everyone else has.

To stop gaps in medical and other needed care for individuals with disabilities, Medicaid should not have to be renewed annually if someone has a permanent medical condition. Medical insurance needs to be permanent too.

We need to stop being fragmented in our advocacy. When organizations name themselves "autism," they exclude all others with developmental disabilities and even typical children. Although people may have different abilities and disabilities, and specific research and interventions need to be tailored to a person's specific diagnosis and needs, our goals are the same: free and appropriate education and appropriate care after the child finishes school. This necessitates individualized treatment. We need adequate funding for our schools, after-school supports, and programs for older individuals. While there is a lot of money to fund war, there is little money to fund children and adults with special needs. If we all join together, we can get a lot more done.

## Is the School Appropriate? Terms to Know

### *Individualized ABA*

Under an individualized ABA, criteria and time intervals for reinforcement depend on the individual child's behavior and cognitive abilities. Preferred reinforcers are assessed for each student.

For example, one student's behavior might be reinforced with a piece of favorite cookie or chip every five minutes for not mouthing, or earning a preferred toy or video clip for every seven tokens earned for correct responses. Another student may be reinforced every seven minutes for not having a tantrum with tokens placed on a ten-token board. A student may earn frequent edible or other preferred reinforcers during acquisition of programs without a token board.

### *Generalized ABA*

Under a generalized ABA, everyone has access to the same reinforcers. The time duration to earn reinforcement is the same for each student. The reinforcement criteria are the same for all students.

For example, everyone in the classroom earns the same tokens every thirty minutes for not disrupting the class, and everyone has the same primary reinforcers (the same food or activities) for earning the same number of tokens. In this case, students do not necessarily understand why they are being reinforced or not being reinforced. The token becomes an object of self-stimulatory behavior that interferes with learning, such as a student perseverating on a paper-bracelet token.

### *Individualized TEACCH*

Visual schedules are based on the individual's cognitive level and designed to motivate a student. Evaluations to assess pictures (i.e., photos or Mayer-Johnson symbols) to object correspondence are completed prior to the schedule. The number of icons (pictures or objects) that a student can discriminate in a field is assessed.

One student may use an object-activity schedule because the student does not understand picture-to-object correspondence, another student has a photo-activity schedule, and a third student uses Mayer-Johnson symbols. If a student likes trains and has picture-to-object correspondence, the schedule is placed on a train background. If a student likes the color purple, the schedule is done on a purple background. If a student cannot discriminate multiple pictures or objects in a field, then the schedule is done in a photo-album format with one icon on a page.

### *Generalized TEACCH*

Visual schedules are the same for all students. It is not clear if the students have prerequisite skills such as two- to three-dimensional matching ability.

For example, all students would have picture-activity schedules using Mayer-Johnson symbols. These are lined up next to each other on the wall, even though some students may not know what the icons represent or may be unable to discriminate with more than three pictures in a field. The schedules are not visible from where students do their work much of the day.

TEACCH is not PECS (Picture Exchange Communication System), both discussed elsewhere in this book. I find it frustrating when I visit placements, and special-education teachers are confusing TEACCH with PECS. The TEACCH program was developed at the University of North Carolina by

Eric Schopler. For further reading on TEACCH, I highly recommend *Clinical Manual for the Treatment of Autism, American Psychiatric Publishing*.[319] I particularly find this paragraph very informative:

> These three structured teaching components can be implemented in all age groups, for all developmental levels. For example, in the home of a preschooler, a parent can designate and set up areas for play, teaching time, and snack time. For a preverbal child, the parent can use a row of objects to indicate the order in which activities will occur, such as a cup for mealtime followed by a ball for play time. For a preschooler who has emerging language and an understanding of pictures, a photo or line drawing schedule of two or three activities in sequence can be posted in a convenient area. Students in elementary school classrooms, whether self-contained or included, can also be provided with these elements of a structured program. Because no two children are exactly the same, a good classroom should reflect a diversity of schedules, work systems, and visually structured tasks.[320]

PECS was developed by Andy Bondy and Lori Frost as an aid and augmentative alternative communication technique. In TEACCH, objects or pictures are used to show an individual a visual schedule to follow. In contrast, with PECS, individuals with impaired verbal skills communicate to others using pictures. Individuals are reinforced for their requests using the pictures, and expressive language is paired with the pictures to develop verbal skills and spontaneous speech often develops.[321]

## Recommendations to Parents

By law, a school district has to provide an appropriate placement according to your child's needs. As a parent, be knowledgeable, know what to look for in a

319 Marcus and Schopler, *Clinical Manual for the Treatment of Autism*, 211-33.

320 Ibid., 221.

321 Fry Williams and Lee Williams, *Effective Programs for Treating Autism Spectrum Disorder: Applied Behavior Analysis Models, 156-57.*

placement, and know what you need to ask for. Your knowledge is your best weapon against a system that might be against your child, and it is something no one can confiscate!

As a starter, I highly recommend the textbook *Effective Programs for Treating Autism Spectrum Disorder: Applied Behavior Analysis Models.*[322] Early chapters explain the technical jargon, making the rest of the book easy to understand. The later chapters discuss what effective programs do to teach children and how they individualize treatment.

After reading this book, I was much better at evaluating placements the DOE recommended for Talia. Spending a few days reviewing some of the chapters before an impartial hearing makes me feel more confident and reduces some of my anxiety. Fortunately, the DOE and I have settled every year, and I never had to be examined or cross-examined at a hearing.

Be organized. It is not easy to do when you are raising a disabled child. I found it helpful to have a large paper calendar to keep track of scheduling. Therapists can write in their time slots, so they do not arrive at the same time. This helps us make sure that Talia is always at home when therapists arrive. I also write a Talia-related "to do" list at the bottom of the calendar.

Do not throw out old records, because you may need them one day. If your child has private therapy, keep careful track of all invoices and canceled checks. Do not write checks without getting signed invoices. Keep them in order. Initially I did not keep them in good order, and two years later, when I received authorization for reimbursement, it entailed hours of tedious work to organize them.

Take careful notes at IEP meetings and proposed placements. Take a list of questions to any proposed placements. In-depth knowledge about your child's condition will help you ask specific questions. Ask about schedules of the day. See if the curriculum is on your child's level. If your child has problem behaviors, ask how the school conducts functional behavior assessments. Ask if they do functional experimental analysis. Ask about their behavior plans, including how and how often reinforcer assessments are done. If your child

---

322 Fry Williams and Lee Williams, Effective Programs for Treating Autism Spectrum Disorder: Applied Behavior Analysis Models.

needs discrete trial training, find out how this will be implemented according to your child's level (whether 1:1 ABA will be done if needed and for which portion of the day). Find out the qualifications of the staff. See what children are doing in the classroom, and if and what they are learning. Ask about the related services. Are they done in or out of the classroom? Does your child need it to be provided in a quiet area because of distraction? If your child has noise sensitivity, pay attention to the noise level. See if ADLs, such as showering and dressing, can be worked on if your child needs them. Of course, make sure you bring this up at your IEP meeting if they are not already draft goals. Ask if staff uses a task analysis in which the activity is broken down into multiple small steps and if physical prompts are used-which, unlike verbal prompts, can be gradually faded out.

I have treated older individuals with autism who clearly had the capacity to be toilet trained but never were. Many schools refuse to remove diapers for toilet-training programming. Even a typical child cannot learn to use the bathroom if the child's diaper is not removed.

I have had female patients who could not go out in to the community or attend a day program when they have their menses, because they refused to wear a sanitary napkin. Once their bodies are grown, it can be very hard to compel them to wear one, much less teach them to change it independently. If they are taught as small children, such as at age nine, it would give them time to learn to tolerate, put on, and change them, so they are ready by the time their menses starts. This would avoid a lot of dysfunction in the future.

For economic reasons, ADLs should be given priority. It would decrease the need for the government to pay people to perform these activities for people with developmental disabilities. We can either be their teachers or their servants. It does not matter how well a person may know history or science, if that person does know how to shower or dress independently.

### *Make Sure Your Child's Educational Treatment Is Research and Evidence Based*

I am sure you would not accept a pediatrician's advice to prescribe garden weeds to your child for an ear infection. Do not allow your child to be in an educational setting that uses a methodology that lacks supporting evidence. Even if the setting uses an evidence-based methodology, make sure it is used the same way it was in research, just as you would want your child to take medicine the same way as it was researched and tested. The Individuals with Disabilities Education Act and No Child Left Behind Act specify that instruction must be scientifically based with peer-reviewed research.

### *Beware of Barefoot Speech Therapists*

To save money, until February 2, 2004, New York State gave certificates to practice speech and language therapy to individuals with only thirty college credits in speech and language disorders. An applicant with a master's degree would receive a permanent certificate, regardless of the subject of the degree. Therapists were required to have only 150 hours of supervision in a school setting, not necessarily hours in a clinic, and two years of paid full-time supervised teaching experience, but no observation hours.[323] Although no new individuals can obtain these certificates, some practitioners still work as speech therapists under the old certificates.

To obtain a speech and language pathology license in New York now, an individual is required to have a master's degree in speech and language pathology and to have at least 375 supervised clinic hours, with 325 of them at the graduate level. The applicant must also undergo twenty-five observation hours and thirty-six week full-time supervised work experience.[324]

To check if someone is a licensed speech and language pathologist, check the National Provider Identifier Database at http://npidb.org/ and search for the therapist's name. You can find Prompts for Restructuring Oral Muscular Phonetic Targets (PROMPT)-trained therapists in your area-these therapists are helpful for problems with verbal apraxia or motor planning-by checking the PROMPT Institute website (www.promptinstitute.com). You can cut and paste therapists' e-mail addresses into messages. If a school tells you their

323 "SLP Credential Comparison, Requirements to Earn Speech-Language Pathology Credentials in New York State," New York State Speech-Language-Hearing Association, accessed February 14, 2016, http://www.nysslha.org/i4a/pages/index.cfm?pageid=3492.

324 Ibid., accessed January 10, 2016.

speech therapists are PROMPT trained, ask for level the therapists are trained at.[325] I once had a DOE-recommended state-approved private school education director tell me that all the speech therapists were PROMPT trained, and when I asked what their level of training was, I was informed that the therapists were seeking training.

If you live in New York City, there are two lists for speech, occupational, and physical therapists: one is for individuals employed by the DOE, and one is for other individuals and agencies. The New York City DOE has a special "district" for most children in special education: District 75. There are District 75 coordinators who work for each related service. They can send an e-mail blast for you. They can screen people who would be appropriate for your child's disability and who can come to your home. The occupational and physical therapy coordinators were helpful. They found a few people who were interested and called me. I found one of the coordinators just by calling someone on the list who shared her name and contact information.

However, if there is a chance you may go to hearing and a therapist may need to testify, it is preferable to use someone who is not employed by the DOE. To obtain lists of therapists, plug "New York City Department of Education Municipality List" in an Internet search. You might also need to add to your search the type of related service, such as physical therapy. There are separate lists for preschool therapists. These are printed by each borough and include all related service providers, including physical, occupational, speech, and counseling. If you live in another part of the country, find out if lists of therapists exist or if there is a coordinator who can send out an e-mail blast for you.

Furthermore, in New York City, if your child goes to a center, you can get your transportation costs reimbursed, even for car services and metered taxis, via the RSA-3 form. If you live outside of New York City, inquire into getting transportation reimbursement for services after school. However, if you are going to try to obtain a monetary stipulation of settlement for private therapy, and the transportation cost is minimal, it may be better not to request to

325 There are four levels of Prompt Training: "PROMPT Training," PROMPT Institute, accessed February 14, 2016, http://www.promptinstitute.com/?page=PROMPTTraining.

obtain the reimbursement for transportation. The DOE may instead subtract more money from the total settlement than the transportation costs. Check beforehand with a special-education attorney.

In recent years, the New York City DOE has contracted out to agencies that have given the lowest bids. There have been instances where these agencies have had no available therapists or the assigned therapist has been untrained to work with the child's specific disability, which I liken to a gynecologist removing an appendix. In some cases, with letters from parents, the CSEs have allowed specific therapists who are not employed by the lowest-bid-contract agencies, but the parent may choose on the basis of the specific training to work with a child's specific disability.

You can pay privately, if you cannot get a therapist, and request reimbursement in an impartial hearing request, but check first with a special-education attorney. You can also ask a special-education attorney to find out if there is a mechanism to choose your own therapist on the municipality list who can receive direct payment from the DOE. It is unfortunate that children of parents without the funds may not receive the services they need, even though the services are on the IEP.

### *Advocate to Therapists, Doctors, and Agencies to Write Recommendations Properly*

Once when the CSE requested an evaluation from Talia's therapists, the director of the agency providing occupational therapy refused to write the recommendation. She stated that other children did not get as much occupational therapy as Talia did, and it did not make sense to send her to the therapy gym so often. But then I gently argued that my daughter needed help with ADLs such as bathing, dressing, and the like. She agreed and wrote a recommendation to continue the therapy at the same frequency as before, with some of those days at home or school, to work on ADLs.

When Talia had significant regression in speech therapy during a school break, I asked her therapists to recommend services for fifty-two weeks a year. Initially, the director of the agency providing the services argued that no such thing existed. I argued my case to her about Talia's significant regression. The therapists recommended continuous therapy, and I was able get it in writing on Talia's IEP through an impartial hearing settlement. If your child needs any service, *do not take no for an answer!*

During an IEP meeting or in a report from your child's therapists, suggest your child's therapists to use correct legal terminology: "appropriate," "needs," or "requires." If you are seeking a residential placement, try to ensure the recommendations specifically state "residential educational" placement. Unlike residential educational programs, standard residential treatment facilities-which are not schools, and the school districts do not have to fund-do not have to provide an education and heavily rely on medication management. If your child has severe behaviors, and you want an ABA residential educational program, suggest your child's treating professionals to write that your child needs a consistent twenty-four-hour behavior plan because of the dangerous behaviors. Suggest your therapists to avoid terminology and recommendations with words such as "best," "would benefit from," or "gold standard," because the government has no legal obligation to provide a service simply on the basis of what would benefit the child or would be the best, and those terms will not hold up in a courtroom. You can suggest at an IEP meeting that the CSE consider something, but "consider" should not be written in the recommendations of a report. Professionals should use stronger words such as "needs" and "requires."

Save all envelopes with postmarks from the CSE, in case deadlines such as meeting notices are missed. For example, these are supposed to be received five days prior to an IEP meeting. If a placement offer is sent without an enclosed IEP (the postage would then be low), this will be your evidence that you received a school recommendation without an IEP.

Do not get angry and certainly do not yell (even though it may be hard not to) at an IEP meeting, just like you would not yell at a police officer. If you do not agree with it, do not sign the IEP. Just tell them you will need to think it over or consider your options. Only sign the attendance sheet, not even the minutes.

Take detailed notes. Note the times the meeting started and ended and get the names and titles of everyone present. Document if any of them leave the room and the time they come back, or if they do not come back. Note if you received reports or evaluations at the meeting itself, as you are supposed to have these before the meeting. The CSE giving them to you at the meeting can prevent you from being able to fully participate in your meeting and can therefore help your case if you need to request an impartial hearing.

Never tell them you will get a lawyer, unless your lawyer tells you otherwise. Threatening to get a lawyer is a message to the CSE that you are preparing actively for litigation and so should they.

Tell them any goals or services you believe your child needs and request those services, because if you do not tell them at the meeting but then later request those services in an impartial hearing, they can be denied. You were supposed to request that at the meeting.

Do not tell them you will not accept a placement or class size they offer without first visiting the program. Being "open-minded" is necessary to winning an impartial hearing. You cannot go to an IEP meeting stating you will not consider other possibilities besides a private school or a home-based program.

Beware of trap questions (like those I was asked), such as this: "Do you plan to keep your child at (blank) school until the age of twenty-one?" An appropriate response may be, "First, I need observe what other placements you offer," or "I have to see how my child is doing."

If you do not agree with the district's IEP and the CSE makes a mistake, and you need to file an impartial hearing to solve the problem, write the CSE's mistakes in the impartial hearing request. The more obvious it is that the CSE violated your child's rights, the better chance you will have to prevail at hearing.

If the school or the CSE suggests medication, remember it is illegal for medication to be part of an IEP. No one should be able to put your child on a medication with potential toxic side effects without your permission. You have a legal right to refuse consent.

Do not assume that just because individuals are professionals, they know best. I submit government employees and even private placement employees may have a legitimate professional opinion but they also have to work under "gag rules," so they may be fired if they recommend an appropriate placement that politicians do not want to fund. They cannot freely recommend services their professional judgment would require. Some may be adamant about insisting children do not need certain services because of the politics involved. Do not listen to these prostitutes of their professions.

Although I have met some school district officials that cared, they were limited in what they could recommend. Officials who care and try to help your child may be unappreciated and even disliked by their supervisors. Make sure you let them know you are grateful.

Be careful about the e-mails you send. All e-mails, except for those exclusively between you and your attorney-which are considered privileged-can be subpoenaed. If you think what you plan to send in an e-mail to anyone except your lawyer could possibly be used against you, do not send it. Give the person a call, instead. Remember, what you write can be taken out of context by the school district's attorney.

Visit all schools the school district formally (and perhaps even informally) refers you to, unless the placement rejects your child by phone or a special-education attorney says otherwise. For example, a special-education attorney might inform you that the number of schools recommended is excessive and that you may not need to see anymore schools after you have already visited a number of possibilities. When you do visit a placement, take detailed notes, just like at an IEP meeting.

Expect progress, not miracles. Make sure your child is getting enough hours of therapy. As stated earlier, a child may need up to forty hours weekly of 1:1 ABA plus supervisory hours.

If a therapist is not working effectively with your child, discuss it with the therapist and offer suggestions. For example, I observed one speech therapist soon after she started with Talia. She would allow Talia to pick up objects and tried to get Talia to say the words. I gently explained that this would be ineffective with Talia, because she needed structure and repetition. I bought some Kaufman praxis cards (which the prior therapist used) and asked her to sit Talia at a table and work with them. She was receptive to my suggestions; she worked with Talia for almost two years.

If a therapist will not discuss your concerns or suggestions or is stubborn about his or her own way of working, without individualizing therapy to your child, then get a new therapist. Discuss with a lawyer about hiring your own therapist and then filing an impartial hearing to seek reimbursement. Do not waste your child's time with babysitting. However, do not deny the problem and think your child may be ready for something too advanced.

Make sure you receive parent training and counseling. Research in parent training has shown the children of trained parents can improve behavioral outcomes with less medication[326] and that the training is more effective compared to educational sessions providing information about autism without behavior management.[327] Training and counseling are mandated for the parents of certain children in New York, including parents of children with autism. It has to be individualized to your child, not some general workshop. If you plan to file for an impartial hearing in New York for other services, do not remind the CSE if parent training and counseling are not on the IEP. Instead, state at the impartial hearing request that you are not receiving the parent training and counseling that should be required on the IEP.

If your child exhibits dangerous behaviors on long bus rides, tell the driver and aide to document all incidents. If the behavior is distracting the driver, have the driver write a letter stating this. The documentation will support a limited time travel on the IEP. If the bus were to have an accident, the school district will have difficulty defending itself when previously forewarned of the driver's impaired ability to focus.

If you need a lawyer, make sure you use someone who does primarily special-education law, not someone who dabbles in it. Special-education law is not generally studied in law school or practiced in general. It requires specific training, and the regulations and case law change constantly.

326 Aman, McDougle, Scahill, Handen, Arnold, Johnson, Stigler, et al., "For the Research Units on Pediatric Psychopharmacology Autism Network: Medication and Parent Training in Children with Pervasive Developmental Disorders and Serious Behavior Problems: Results From a Randomized Clinical Trial," 1143-54.

327 Bearss, Johnson, Smith, Lecavalier, Swiezy, Aman, McAdam, et al., "Effect of Parent Training vs Parent Education on Behavioral Problems in Children with Autism Spectrum Disorder, a Randomized Clinical Trial," 1524-33.

Send all correspondence to the lawyer. Do not send any progress reports or testing to the district without sending it to your lawyer first. School districts will twist things around and take things out of context. Do not sign a tuition contract without a lawyer first reviewing it, or you can end up losing your case at an impartial hearing just because of the contract wording. For example, if the contract has a nonrefundable tuition deposit, if you pay it, and the district recommends a placement before the school year starts, an impartial hearing officer may deny you tuition reimbursement: you were "close minded" to accepting any district placement, because you already paid a nonrefundable deposit, that is, you were looking for reasons to reject the school district's recommendations.

When your child reaches the age of eighteen, you must obtain a legal guardianship: this is not automatic. With a legal guardianship, you have a right to consent to treatment, including medications. If your child is hospitalized, let the staff know right away that you have a legal guardianship and only you are authorized to provide consent. If they give your child medication without your consent, contact the risk-management office and send over the legal guardianship papers. For other problems, you can also contact the patient advocate's office.

Be wary of alternative treatments presented as cures. People will sell you the Brooklyn Bridge if they can convince you this will help with your child's autism.

Make sure your children are receiving physical exercise, which can help with autistic behaviors as well as academics.

Find balance and meaning. The Friday after Thanksgiving in 2011, I got on the scale and found I weighed 131 pounds. I am only five-foot-one, and I was disgusted with myself. Furthermore, my blood sugar for the prior two years showed I was at risk for diabetes.

My husband bought an elliptical machine for $150 on a Black Friday deal. After I used it for ten minutes, I was exhausted-but I did not give up. Gradually increasing my time, I started to use it thirty-five to forty minutes a day. Then I started measuring out my food, such as a cup of cereal, and realized I had been taking much more than a standard serving size.

I made myself a measurable objective (like an IEP is supposed to have) to get to 105 pounds by my birthday in October. I had wanted to lose weight in the past, but now I found when I made my goal measurable with increments of one pound a week, I had enough motivation. I later took up running.

I made myself another goal to accomplish by my birthday, which was to have forty thousand words for this book, writing five hundred words a week, and I made that goal too. Making these goals helped me to set priorities and do fewer other things that I really did not have to do, or other people could do. But writing this book was something only I could do.

Since I turned twenty-one, birthdays have, for the most part, been upsetting. However, now (in my forties) my birthday felt good, because I saw I had accomplished so much in the past year.

I continued running and have been running eight-minute miles. I won a second-place medal for my age group in a five-mile run less than a year after I started running. I was especially surprised because as a child I was bullied in gym (I would drop balls and was a slow runner). I dreaded going to gym. At the time, my mother explained the problem to the doctor and then asked him to write a note stating I had asthma, which got me excused (even though I never had asthma). About eighteen months after that race, on October 26, 2014, my husband and I ran the Cape Cod Marathon, in which we raised $4,685 for the JRC. My time was 4:32:10 with an average mile of 10:23, and I beat my husband by almost one hour. At my school district, I have been referred to as the parent who jogs to the CSE. In addition, on November 15, 2015, I completed the Brooklyn Marathon. My time was 4:26:33 with an average mile of 10:10. I am now training to try to qualify for the Boston Marathon. With running, weight lifting, and abdominal exercises, I am no longer at risk for diabetes.

If your loved one with a disability has siblings, consider having them join a siblings' group. Batsheva enjoys bonding with other children who also have a special-needs sibling. Do not forget to have private time with your other children.

One day while my mother watched Talia, I took Batsheva into Manhattan for a tour of Grand Central Station. On the spur of the moment, I suggested we find out if we could get half-price tickets to a Broadway show. After waiting almost an hour on line, I found out that even at half price, a show was seventy-two dollars a ticket. I stepped out of the line to call my husband to ask if it was okay to spend so much. He told me to go. We had fifth-row center-orchestra seats. My daughter and I had a great time, but I later felt guilty about spending so much. Both my husband and mother told me I should not feel that way, because it was important to spend time with Batsheva.

Most of all, never be resigned and submissive when it comes to your child. Do not allow the government, a school, day program, or residence to be a steamroller and yourself the pavement.

# Concluding Notes

## A Tribute to My Father, Myron Slaff, Mayer ben Yisrael

I write this three days after he passed away, on July 23, 2015.

When I was a child, we would take long walks on weekends, and he would always stop at the store and buy me a lollipop. He used to enjoy taking me out to lunch and giving me ice cream, and he made for me and my friend the most delicious ice cream sodas.

When I was older and asked for help to pay for Talia's therapy, he would give me money without reservation. When I received a reimbursement check from the DOE and I wanted to give him back his money, he would tell me to keep it, that I would need it again.

When I went years without a vacation, he offered me $2,000 and told me to go away with my husband and have a good time. He and my mother watched the children so my husband and I could get away.

When he thought buying a used car could be unsafe because it might not be maintained properly, he bought me a new car. I drive that car today.

When Talia had no school bus, and I had to work, he took her back and forth to and from school every day by taxi. He stayed at the school or went to the nearby library for the day, because it was too far to go back to my house and return to the school with public transportation. He took her to therapy after school.

He took his mother-in-law to her doctor's appointments. He let my orphaned cousin live in our home and fed her without any compensation. He always gave money to beggars.

I also realize that my father was on the autism spectrum. He was a supervisor accountant for New York State. He had received the highest score of all the examinees on a civil service exam.

He always had fixed routines. Dinner was at eight o'clock. I liked the vacation routine we sometimes had: ice cream at four o'clock. I also enjoyed the melons, strawberries, and blueberries he brought over and prepared every time he came for dinner. A few minutes before he passed, I spoon fed him some fruit cocktail that I remembered he always enjoyed sharing with me when I was a little girl.

The only books he read were a prayer book and books about airplanes. He collected useless items. He could not part with old things.

He was always distant, and I could not relate well to him. For the most part, I could not find comfort in discussing my feelings or problems with him. He was blatantly honest, which hurt my feelings. My last gift to him was a book about airplanes. When I asked him how he liked it, he told me that some of the pictures were good but that it got boring after a while.

His distance made me angry and resentful. I said some mean things, but now I realize that, just like my brothers and my daughter, it was not his choice to be on the autism spectrum, and I forgive him for being so distant. I hope he forgives me for some of the mean things I said. I know he really loved and cared for his family.

People do not choose to have autism or any other disability, but we can choose to love and take care of them.

# Appendix

## History of Effective Treatments Used in Autism

In this section, I discuss evidence-based treatments that have been talked about elsewhere in this book.

### *Applied Behavior Analysis (ABA)*

Behavioral psychology originated with John B. Watson, PhD, who published in 1913 "Psychology as the Behaviorist Views It."[328] He believed that the relationships between environmental stimuli and behavioral responses could be objectively measured unlike in other schools of psychology. B. F. Skinner, PhD further developed this science by examining frequencies of future behavior after changes in the environment contingent on previous behavior. This was known as operant conditioning where the future frequencies of responses could change depending on the consequences of past responses.[329] Today, in applied behavior analysis settings, ABC (Antecedent Behavior Consequence) data are always taken, stating the antecedent (what happened in the environment, who was there, and where and when it happened), the behavior itself, and the consequence

328 John B. Watson, "Psychology as the Behaviorist Views It," *Psychological Review* 20, (1913): 158-77.

329 Sadock, Alcott Sadock, Ruiz, *Kaplan and Sadock's Synopsis of Psychiatry*, 11th ed., 101.

(what happened after the behavior occurred). These data are used to examine why the behavior occurred, known as the function of the behavior. For example, if the consequence of stealing food off others' plates is eating the food and that food is liked by the individual, that behavior is likely to occur again. We can then change the frequency of the behavior by giving more desired food items when not engaging in the behavior for a predetermined length of time. Using positive reinforcement can increase the frequency of future desired responses.

In 1949, Paul R. Fuller taught an individual with profound intellectual disability to move his right arm to receive an edible reinforcer, a sugar-milk solution.[330] In the 1950s and 1960s applied behavior analysis further showed that individuals with profound intellectual disabilities were capable of learning and not hopeless as previously thought. "Over 1000 peer-reviewed, scientific autism articles describe ABA successes."[331]

O. Ivar Lovaas, PhD, in 1987 showed that children with autism under forty-six months of age receiving forty hours a week of 1:1 behavior intervention for two years has a mean IQ gain of thirty points compared to a control group.[332] Furthermore, almost half the children receiving the intensive 1:1 applied behavior analysis intervention passed first grade. This study utilized mostly positive reinforcement but also utilized aversives. A follow-up study done when the children reached an average age of 11.5 years showed that the children maintained their IQ gains and also had improvements in adaptive behavior compared to the control group.[333]

### *Aversive Therapy*

As stated earlier, aversives are procedures that are arranged as consequences for undesired behaviors with the purpose of decreasing the future frequency of those behaviors. Some aversives, such as bad grades, money fines, or critical

330 Paul R. Fuller, "Operant Conditioning of a Vegetative Human Organism," *American Journal of Psychology* 62, (1949): 587-90, doi:10.2307/1418565.

331 Foxx, "Applied Behavior Analysis Treatment of Autism: The State of the Art," 822.

332 Lovaas, "Behavioral Treatment and Normal Educational and Intellectual Functioning in Young Autistic Children," 3-9.

333 John J. McEachin, Tristram Smith, and O. Ivar Lovaas, "Long-Term Outcome for Children with Autism Who Received Early Intensive Behavioral Treatment," *American Journal on Mental Retardation* 97, no. 4 (1993): 359-72.

comments, do not involve discomfort or pain. Those aversives are often very effective with most people and are widely used and accepted procedures. Unfortunately, they are sometimes ineffective with individuals who have autism. For these individuals, effective aversives may have to include an element of discomfort or pain, such as a spray of water mist to the face, a spank, or a pinch, in order to be effective. Aversives have been used since ancient times. Aversives have been used inappropriately and unethically in the past for behaviors such as homosexuality, but some researched applications have shown positive results in alcoholism, thumb sucking, and electric shock for self-injurious behavior for individuals with intellectual disabilities.[334] For thumb sucking, a bitter fluid is placed on the fingernail. Such fluids are widely available for sale at pharmacies. Furthermore, Antabuse (disulfiram) is an FDA-approved medication for alcohol dependence, and it is used as an aversive by producing unpleasant symptoms after alcohol intake.[335] These aversive symptoms include "nausea, throbbing headache, vomiting, hypertension, flushing, sweating, thirst, dyspnea [difficulty breathing], tachycardia [rapid heart rate], chest pain, vertigo [feeling that one's body is spinning], and blurred vision."[336] The majority of studies on aversion therapy have used skin-shock treatment. "[P] unishment may be critical to treatment success when the variables maintaining problem behavior cannot be identified or controlled."[337] Indeed, positive-only behavioral interventions are not always effective.[338] Specific limitations are discussed further elsewhere in this book. There are currently 119 peer-reviewed articles supporting the use of skin-shock treatment.[339]

Negative reinforcement, not to be confused with aversives or punishment, is the removal of a stimulus after a response that increases the likelihood of that response reoccurring. For example, if an individual, after performing a task correctly, receives a break from work, then that individual is likely to

334 Council on Scientific Affairs. Council Report. "Aversion Therapy," 2562-66.

335 Sadock, Alcott Sadock, Ruiz, *Kaplan and Sadock's Synopsis of Psychiatry*, 11th ed., 966-67.

336 Ibid., 967.

337 Lerman and Vorndran, "On the Status of Knowledge for Using Punishment: Implications for Treating Behavior Disorders," 432.

338 Carr, Horner, Turnbull, Marquis, McLaughlin, McAtee, Smith, et al., "Positive Behavior Support for People with Developmental Disabilities: A Research Synthesis," 45.

339 Bibliography on Skin-Shock, accessed January 24, 2016, http://www.effectivetreatment.org.

perform the task correctly again. The response of completing the task correctly increases in frequency. Therefore, the break from work is providing negative reinforcement.

Ogden R. Lindsley, PhD, one of Skinner's students, introduced the terms "accelerating consequence" and "decelerating consequence."[340] Any event that causes a behavior to increase in frequency is an accelerating consequence. Any event that causes a behavior to decrease in frequency is a decelerating consequence. This is a more accurate description of the use of an aversive rather than referring to an aversive as a form of punishment.

### *Picture Exchange Communication System (PECS)*

PECS was invented by Andy Bondy, PhD, and Lori Frost, MS, CCC/SLP, for individuals to communicate using pictures. The program teaches individuals to hand pictures to others to communicate their needs and later on to combine the pictures together in sentences.[341] Using electronic devices, communication programs such as Proloquo2go combine verbal language with pictures and written words to facilitate independent expressive language.

### *Prompts for Restructuring Oral Muscular Phonetic Targets (PROMPT)*

PROMPT is a researched-based speech and language therapeutic intervention used in various conditions including autism.[342] It has been developed over thirty years "as a treatment for speech production disorders in both children and adults based in accepted neuromotor principles of speech production."[343]

---

340 Ogden R. Lindsley, "Theoretical Basis of Behavior Modification," *Kansas University, Lawrence, Bureau of Child Research*, May 1967, 5.

341 Fry Williams and Lee Williams, *Effective Programs for Treating Autism Spectrum Disorder: Applied Behavior Analysis Models*, 156-57.

342 Sally J. Rodgers, Deborah Hayden, Susan Hepburn, Renee Charlifue-Smith, Terry Hall, and Athena Hayes, "Teaching Young Nonverbal Children with Autism Useful Speech: A Pilot Study of the Denver Model and PROMPT Interventions," *Journal of Autism and Developmental Disorders* 36, no. 8 (2006): 1007-24, doi:10.1007/s10803-006-0142-x.

343 Ibid., 1009.

In individuals with autism, when the child produces "an intentional sound to request…[t]he child's utterance is then supported through integrated and auditory tactile cues. The adult uses both vocal modeling and actual manual manipulation of the child's jaw, lips and other speech mechanisms while the child vocalizes to elicit speech approximation of a target word. Physical cues are gradually faded into visual cues, so that the child responds to a hand movement rather than a touch, and then further faded."[344] Similar to ABA, prompts are gradually faded in PROMPT speech and language therapy. Later, "syllables are shaped into words and short phrases."[345] The goal is to speak independently. There are different levels of PROMPT training as well as certification, each with its own requirements.[346]

### *Treatment and Education of Autistic and Related Communication Handicapped Children (TEACCH)*

TEACCH was developed by Eric Schopler, PhD, whose doctoral dissertation in 1966 showed that "autism and related developmental disorders were primarily impaired ways of experiencing the world and understanding that experience. This included impaired and unusual sensory processes, unusual ways of thinking and understanding, restricted social interactions, impaired communication, and social interests."[347] He noticed that many children with autism would touch and smell objects but not pay attention to what they saw or heard. The study just as importantly showed that autism was not due to problems with parenting. Most importantly, TEACCH research showed that impairments in autism could be improved with education in the individuals with autism and by training their parents.[348] In the 1970s, TEACCH educational

---

344 Ibid., 1010.

345 Ibid.

346 "Find a PROMPT SLP," PROMPT Institute, http://www.promptinstitute.com/search/custom.asp?id=3790.

347 Gary B. Mesibov, Victoria Shea, Eric Schopler, Lynn Adams, Elif Merkler, Sloane Burgess, Matt Mosconi, et al., *The TEACCH Approach to Autism Spectrum Disorders* (New York, NY: Springer Science and Business Media, 2004), 2.

348 Ibid., 3.

methods were developed.[349] Classrooms utilizing TEACCH methodology were some of first educational placements in the United States for children with autism.[350] The majority of individuals with autism were found to learn better visually in comparison with auditory means.[351] Research also discovered that individuals with autism responded better to a structured environment.[352] Individual daily schedules and routines were also utilized.[353] The TEACCH program also notes the variations between people with autism and the need to assess each individual to determine their unique needs.[354] Although TEACCH methodology assesses deficits, TEACCH methodology also recognizes skills.[355] Services were also expanded throughout the life span,[356] and there has been a focus on community integration.[357] The TEACCH program focuses on visual stimuli for learning and fostering independence.[358] Physical structure includes organizing the furniture and establishing boundaries for the student as well as minimizing distractions to facilitate learning.[359] Research has shown that using a physically structured environment can help with behavior and foster academic gains.[360] However, the physically structured environment must be

---

349 Ibid., 3-4.

350 Marcus and Schopler, "Educational Approaches for Autism-TEACCH," in *Clinical Manual for the Treatment of Autism*, 212.

351 Mesibov, Shea, Schopler, Adams, Merkler, Burgess, Mosconi, et al., *The TEACCH Approach to Autism Spectrum Disorders*, 3.

352 Ibid., 4.

353 Marcus and Schopler, "Educational Approaches for Autism-TEACCH," in *Clinical Manual for the Treatment of Autism*, 213.

354 Mesibov, Shea, Schopler, Adams, Merkler, Burgess, Mosconi, et al., *The TEACCH Approach to Autism Spectrum Disorders*, 8.

355 Marcus and Schopler, "Educational Approaches for Autism-TEACCH," in *Clinical Manual for the Treatment of Autism*, 216.

356 Mesibov, Shea, Schopler, Adams, Merkler, Burgess, Mosconi, et al., *The TEACCH Approach to Autism Spectrum Disorders*, 10.

357 Ibid., 11.

358 Ibid., 25.

359 Marcus and Schopler, "Educational Approaches for Autism-TEACCH," in *Clinical Manual for the Treatment of Autism*, 220.

360 Mesibov, Shea, Schopler, Adams, Merkler, Burgess, Mosconi, et al., *The TEACCH Approach to Autism Spectrum Disorders*, 35.

individualized to the specific needs of each child.[361] For example, symbols, photographs, or objects are used for the individual's visual schedules depending on each person's ability to understand the meaning of the type of schedule used. Furthermore, activities needed to be individualized, and the program should be extended into the home for continuity of care.

361 Marcus and Schopler, "Educational Approaches for Autism-TEACCH," in *Clinical Manual for the Treatment of Autism*, 221.

# Index

Made in the USA
San Bernardino, CA
09 November 2018